FIRE HEART

The Life and Teachings of Maya Medicine Woman

Miss Beatrice Torres Waight

In her own words

As Told to Katherine Silva

Edited by Katherine Silva and Ann Drucker

Foreward by Rosemary Gladstar

Copyright © 2011 Katherine Silva. All rights reserved. No part of this book may be reproduced or transmitted in any form or by any means without written permission of Katherine Silva.

Copy Edit donated by Terra Rafael www.wisewomanhood.webs.com and Carol Pancsofar Pierson

Cover design by Katherine Silva & Cover photo by Janet Caspers

Profits from this book are donated to support the education of Maya youth in Belize, Central America

ISBN 978-0-615-48715-1

I dedicate this book to my grandmother, Delfina Tun, my father, Alejandro Torres, my mother, Dominga Torres and all of my nine children~ ***Miss Beatrice Torres Waight***

Miss Beatrice with her Mother, Dominga Torres and her daughter, Zena Waight at home in Santa Familia (photo by Katherine Silva)

Acknowledgments

I give my thanks to Rosemary Gladstar, Ann Drucker and Katherine, Antonio and Matoska Silva. Without their help we could not have written this book which I have wanted to write for many years. ***Miss Beatrice Torres Waight 2/ 2009***

I offer my unending gratitude to Miss Beatrice Torres Waight for opening her heart, her wisdom and her home to me over the past decade. I thank her for entrusting me with the sacred task of writing this book. It has been an immense pleasure.

I give love and gratitude to my husband, Antonio Rey Silva who devotedly supported me in every way through this project.

I send my gratitude to my parents for believing in me and for their love, acceptance and support.

Thank you to my precious son, Matoska for being patient with the many hours his mother was glued to the computer over the past two and a half years.

I appreciate my amazing daughter, Theresa who held the space and gave the tips that got me through the last editing hurdle.

I am very grateful and appreciative to Miss Beatrice's daughter, Marlyn Waight for her strong, joyful spirit and for her help in organizing visits and communications with Miss Beatrice over the past ten years and for answering endless questions for Miss Beatrice regarding this book via email. I thank her also for editing and improving the

Maya Recipes chapter and the Journey to The White Light chapter of this book.

Many thanks goes to Miss Beatrice's daughter, Jeanette Waight for sharing her research regarding her ancestor's history and migration to Belize which provided some important missing dates and details to Miss Beatrice's narration.

Thank you and a big hug to Miss Beatrice's daughter, Zena Waight for sharing her research on the Hetz Mec ceremony which helped clarify for me Miss Beatrice's narration on the subject.

I offer my gratitude and appreciation to Mountain Sun Donna Diamond, for her friendship and her pioneering spirit. It is she who introduced me to the world of herbs and to Belize.

Many thanks I send to Janet Caspers for her help with the plant allies sections in this book and for her wonderful photos and encouragement.

Thank you to Monica Juchum for welcoming Ann and me in her home during hours and hours of editing, raucous gales of laughter, billowing copal smoke, poignant tears and endless cups of osha tea.

Thank you to Terra Rafael for being my book midwife. Her support, patience and assistance with publishing options, copy editing and writer's freak out were invaluable.

Thanks to Joy Kemna for her gentle spirit, her determined research on the roots of Mayan words and her encouragement and prayers.

Thank you to Neta Marlin, Carol Pancsofar Pierson and Ana Johnson for proof reading.

Thank you to Ellen Williamson for her assistance in Spanish spellings and in typing Spanish accents over bubble tea.

Thank you to all of Miss Beatrice's friends who contributed photos and stories for the Memoirs chapter of this book.

I am deeply grateful to Dr. Rosita Arvigo whose inspiring and extensive work in the field of Traditional Maya Medicine in Belize lead me to my own life's work and opened the door for so many of us to experience Maya healing and to meet Miss Beatrice. I thank Rosita from the bottom of my heart for supporting Miss Beatrice in reclaiming her Maya traditions of healing and ceremony even more fully and providing the space that gave her the exposure she needed to become a world travelling teacher and healer. I am very grateful for the sacred container Rosita still holds, The Arvigo Institute, which preserves, protects and promotes ancient Maya healing techniques. I thank you, Rosita, for your advice, your encouragement and for your aid with the scientific names of plants and the spelling of Maya and Spanish words.

I offer my gratitude to Rosemary Gladstar for writing the foreward to this book and encouraging us all along the way!

It has been my immense pleasure to work with Ann Drucker on this project. She is an amazing woman, teacher, healer and dear friend who brings me home again and again. Thank you Ann!

My humble honor and gratitude extends to the beneficial Maya Spirits and Miss Beatrice's Maya ancestors who continue to heal, guide and comfort us! *Katherine Silva 1/11/2011*

I give deep gratitude for Flame Dineen, who supported all of my sojourns to Belize and loved and welcomed Miss Beatrice into our home many times. I give a big thank you to my sweetness parents, Margo and Stanley Drucker, who unconditionally love me with all their hearts. To my sister, Karen, thank you for coming on several trips to Belize with me. Your fun, adventurous spirit made it a truly wonderful experience for my children! To Marian Rose, my spirit sister, who invited me to join her first journey to Belize, introduced me to Miss Beatrice, and also traveled with me on our last trip to say good bye and thank you to our treasured teacher and friend. I give an endless appreciation to Katherine Silva. Your dedication and commitment in both recording and transcribing Miss Beatrice's vast Maya wisdom is now an invaluable gift to us all. Thanks for being my friend!

Ann Drucker

Disclaimer

The teachings, suggestions and recipes in this book are for educational and historical purposes only. Neither the author, transcriber, the editor nor any contributors are recommending these treatments nor claiming them to be safe or advisable, and the ideas in this book are in no way a replacement for medical training or medical care.

Foreword by Rosemary Gladstar

What an honor it is for me to write the foreward to Miss Beatrice's long awaited book. This is such a deeply authentic story, heartfully rendered, of a gentle powerful woman who has touched the lives of so many people.

What makes this story even more moving is that it is told by Miss Beatrice herself, in her own words, as if she were sitting beside you on her doorstep or in her special healing hut, revealing her tale slowly, over a cup of aromatic herbal tea. The story goes on for many days. One can almost hear the river running in the background, and the wind stirring the leaves of the nearby trees, as she talks quietly, traveling back through time, sharing her earthly wisdom.

Born from a long line of healers, shamans, herbalists and midwives, Miss Beatrice came into the world marked with the special birth caul of a healer and a magical tooth that only her mother knew of. Her family, Maya descendents, had traveled from their original home in the Yucatan Peninsula, Mexico over several hundred miles to settle in the small village of Santa Familia, Belize, where Miss Beatrice was born and raised, and where she still lives, not far from the house she was born in. "This is my paradise. It's not that there aren't other places more beautiful, but this is what I know. This is what I love and I think I'm going to die right here. Until then, I'm going to continue to do my healing practice." ~ Miss Beatrice.

Life was not easy. She came from a large family and her parents were typically poor. She says when speaking of her childhood, "There was no recess for me." She started work early in life, worked hard, was married and had several children by the time most young women are still finishing school. But she speaks of the good lessons she learned from those early hardships: how to be strong and to keep on with your dreams no matter what. Though her life was not one of ease, her story is that of "giving ease." From an early age, Miss Beatrice was dedicated to helping others and following a healer's path. And no matter what hardships or challenges, she stayed to the course that was her destiny.

I first had the privilege of meeting Miss Beatrice at her home in Belize when we traveled there on one of our Plant Lover's Journeys. Though it was many years ago, I still recall vividly that first meeting. Dr. Rosita Arvigo, who was our teacher and hostess at Ix Chel farm, where our group was staying, told us about this wonderful healer and herbalist that she wanted us to meet and one afternoon she took us to Miss Beatrice's home. At first meeting, I was so taken by this woman's simple grace, warm kindness, and the powerful healing energy that emanated from her. She was quite simply a kind and radiant human being, a flower in her own garden.

We spent the entire afternoon with Miss Beatrice in her gardens, sharing stories, drinking tea, and picnicking. While we were there, she shared with us her long time dream of a "healer's hut," a simple structure where she could do healing work away from the busyness of her home. After all, along with a busy healer's practice, Miss Beatrice had a house full of children and a husband to take care of. All of us in our group felt so compelled to help her create her dream that we put together a fund that enabled her healer's hut to be built within a year. How it delights me to know that many years later it still serves her and her community so well.

When she came to the United States for the first time, Miss Beatrice was invited along with Miss Hortense, another Belizean elder and midwife, to be featured speakers at the annual New England Women's Herbal Conference. What a special time it was for everyone; it was wonderful to see the way that the participants at the conference were taken by the grace and warmth of these two powerful healers and the teachings they offered. I think for Miss Beatrice and Miss Hortense as well, the event was transformative and allowed them to experience the powerful effect their teachings had on others.

After the conference, I had the honor of bringing Miss Beatrice, Miss Hortense and Dr. Rosita to my home at Sage Mountain and we spent several wonderful leisurely days together. What a treat that was! Miss Hortense and Miss Beatrice would spend hours wandering through the beautiful Sage Mountain Gardens, and like small delighted children come in with pockets full of seeds. I'd see them at night, talking excitedly in their room, busily sorting and storing these seeds for their trip home. I've often wondered how those seeds from these far northern gardens fared in the tropical gardens of Belize.

Since that first trip to the Women's Herbal Conference, Miss Beatrice has traveled to the U.S many times and her teachings have touched the hearts of hundreds of people. She is a rather extraordinary individual: one who embraces a simple life style, but whose wisdom and teachings go deep. As Miss Beatrice says in the opening pages of her story, there are many books written about the Maya people and their traditions, but what makes this one so unique is that it is written ***by*** a Maya woman. Miss Beatrice opens up her heart and generously shares the ceremonies and herbal recipes that she grew up with and now uses in her own healing work. And she shares the traditions of her people, not only from

her personal experience, but from an ancient lineage of healers. Her power runs deep. The story is rich, the recipes abundant, and the teachings generous. Like a fine pot of *caldo cum* (pumpkin soup) cooked slowly over an open flame, there is much to savor here.
Rosemary Gladstar
Herbalist and author

"I am the one, Beatrice, walking through the hills and valleys by the creek, by the river collecting plants for people to heal their diseases. With all my heart I believe it is going to be so. I give thanks to the spirit of the plants in the name of the Father, the son, and the Holy Spirit. Amen.".... Miss Beatrice's Belizean gathering prayer

Table of Contents

Preface by Katherine Silva

Katherine with Miss Beatrice in Belize 2001

"My hope is that by way of this book, the Belizean Yucateca Maya way of healing and living and the inspiring life story and teachings of Miss Beatrice Torres Waight can be passed down for many generations to come."

When Miss Beatrice asked me to help her write a book, I said, "Please give the honor to someone else. I am a very busy midwife's apprentice and a mother. I simply don't have time for this and furthermore, I am not a writer."

Miss Beatrice has a very determined spirit and she did not give up on me. For years she asked me to help her share her message with the world and I finally agreed. I am very glad that I did. Once again, she had seen something in me that I had not yet seen! Through the process of writing this book I have come to adore writing, have been deeply moved by Miss Beatrice's story and have been profoundly touched and transformed by her medicine and her lineage.

It has been my honor to receive, transcribe and edit Miss Beatrice's story and to be her student, friend and client over the past ten years. Sitting in her kitchen in Santa Familia, asking her questions and recording her words, was a precious experience. During the whole process, I was fed so well on Maya home cooking that I must have happily gained ten pounds.

I spent an entire year playing back the recordings and pecking her colorful words into my computer from Sedona, Arizona. I called and emailed Marlyn and Miss Beatrice regularly to clarify what she had said, the pronunciation of people's names and the spelling of Maya and Spanish words.

I returned to Belize a year later to read to Miss Beatrice, her own words, which I had so painstakingly transcribed. It was total fulfillment for me to sit once again in her cozy kitchen, listening to the sounds of the jungle, sipping hibiscus tea and reading to her while she called out, as if in a gospel service, "That's right!" and "Uh huh." She added so much on this second pass, and I became deeply engrossed in her story. By the look on Miss Beatrice's face, I believe she was very satisfied with the book.

Her daughter Marlyn told me, "This book is making my mother very happy!"

Miss Beatrice simply said, "You did a good job."

While having a little siesta, rocking in a hammock in between the book writing sessions with Miss Beatrice, I felt a gentle hand resting on my head and I heard the words, "Thank you."

I opened my eyes and there was no one to be seen. Little did I know this was only the beginning of the miraculous occurrences that would happen as a result of saying yes to this project!

On my last day at her home for this writing session, I had big plans to get all my final questions answered, yet when I walked through her kitchen door with the early morning tropical light pouring in, Miss Beatrice looked at me sternly and said, "No more book!"

"No more book?" I repeated back to her, crestfallen. I didn't understand. I had come all this way. We were almost done. What could she possibly mean?

She laughed and said, "Today we make tamales!"

I joined Miss Beatrice and many of her family members in making tamales of several kinds and bols (a special Maya variation of a tamale) for a special holiday that was coming up the very next day. Her daughters gathered around the fire heart (the fire pit at the center of the home compound) stirring the giant cooking pot full of homegrown corn and laughing together. Miss Beatrice's youngest daughter, Zena was collecting huge plantain leaves from the jungle with her machete for wrapping the tamales. Chickens were running to and fro, tropical birds sang and the immense greenness of the jungle swayed in the breeze. The whole family cracked jokes and chuckled and got the job done. I learned how to make bols and tamales while I sat with Miss Beatrice as she rolled tamale after tamale all day long! What a blessing and a privilege to be interwoven into her family and her culture. The surprise tamale day was a profound teaching for me in letting go, letting the magic happen and in putting family and nourishment first before intellectual work.

The entire ten days I spent in her home that February was a time of sinking into myself, resting, dreaming, taking herbal baths, participating in ceremonies, enjoying the powerful, vibrant, confident, proud, jovial, jesting, feminine energy of Miss Beatrice and all her daughters and allowing the jungle to lure me into timelessness. I will never, ever forget it.

When I first met Miss Beatrice about ten years ago, I was struck by her charming sense of humor and vast knowledge on many subjects. As we sat eating garanachas (a Belizean snack made with beans, cheese, onions and tomatoes on a crunchy corn tortilla) the first time we met, she made me laugh so hard by telling me that if she eats too many chiles she has to put her butt in the refrigerator. She said this while acting out the scene, waving her arms wildly, letting out galls of laughter and invoking chuckles in all who gathered around to listen.

No matter what hardships she experiences, Miss Beatrice finds some way to see the bright side and even crack a joke or two to lighten the load. Miss Beatrice stays strong in the face of adversity and keeps working toward her goals with faith. Her faith and devotion to her Maya ways is very deep and inspiring.

Miss Beatrice has always been a comforting presence to me and has continuously welcomed me with open arms into her home and into her heart. When she touches my wrist with her rue, her basil, her rosemary, her marigolds and white roses and begins to whisper her prayers, it is as if a cool mist of sweet nectarine white light showers over me, cleansing me and restoring me to wholeness.

Miss Beatrice has a huge heart and a strong shining will. In a cultural atmosphere that discouraged women from leading ceremonies and practicing spiritual healing, she has boldly carried the torch of both the Maya women's ways of midwifery, herbs, womb massage and natural well woman care and the previously male roles of Maya spiritual healer and ceremony leader for her family, her village and hundreds of people all over the world. She has been a vehicle for change regarding women's power and women's roles in her family and has raised girls who believe in themselves and who are proud of their Maya heritage. Miss Beatrice is not only a guardian of the Maya way of healing for women, a traditional Maya spiritual healer and leader of ceremonies, she is a loving mother, sister, aunt, grandmother and friend who constantly holds a loving space for her extended family. Her home is a place of joy. Her family shares a connection, love and community of caring and laughter that surpasses any I have previously experienced. I am very blessed to know her.

This book was told to me, in conversation style, sitting at the kitchen table, reclining on the couch, surrounded by the sounds and smells of tortillas being shaped, patted, cooked and eaten with escebeche (sour soup). During Miss Beatrice's many hours of storytelling, her daughters would ask her daily life questions and Bachata music and jungle bird song wafted through the window. While you listen to this story of an incredibly radical and righteous indigenous woman, you may be struck by the intense hardships that she and her people have endured, the tenuousness of life on earth that was her reality and by the immense strength and courage and surprisingly positive attitude she carries. You may be touched by her determination to become the healer she was meant to be in a world that was full of struggle and with teachers who one had to prove oneself to in almost heroic proportions. You may be amazed that this poverty stricken little girl in a tiny village in the jungle who speaks only Maya, who could see fairies and spirits and who followed her father, the village healer and ceremony leader, around trying to grasp and understand the world of medicinal plants, ceremony and healing as he told her that only one of his sons would be the recipient of this knowledge, who toiled from morning until night attending to household chores in her own and many neighboring houses, became the traditional healer for her village as well as an internationally renowned and beloved healer and teacher, traveling to many countries to bring hope, healing and connection to Maya medicine and Maya spirituality.

This book is best read aloud in the company of friends and family, with delicious food nearby, some aromatic herbal tea to sip and a crackling fire.

However you choose to savor Miss Beatrice's delicious words, I invite you to open your heart and your mind to the immense magic of the forest, the depth of indigenous Maya wisdom and the certainty of the strength and power of women everywhere.

This book barely scratches the surface of the immense depth of wisdom and the precious gift that Miss Beatrice carries, but at least it is a start and surely it contains the essence of her wit, her wisdom and her enormous heart.

It has truly been a challenge to translate an oral tradition into a literary form, but Ann and I have done our best to maintain Miss Beatrice's Belizean style, charm and hilarious sayings throughout our editing process.

As she requested, we "changed a little word here and changed a little word there to make it easier to read."

We also found it necessary to organize the narration into chapters and categories because it was told, true to traditional healer style, in a circle, not a line. We changed none of the content and meaning of her actual narration. We did add some facts, dates, definitions, translations and explanations when needed for clarity, provided mostly by her daughters. Also, some of the plant allies information was taken from notes taken by her students in her classes over the years. With the exception of the minor additions mentioned, her book is a direct transmission of her message to the world, in her own words.

My hope is that by way of this book, the Belizean Yucateca Maya way of healing and living and the inspiring life story and teachings of Miss Beatrice Torres Waight will be passed down for many generations to come.

Introduction by Miss Beatrice

I have intended to write this book for a long time! I am so happy that it is finally done! This book contains my life story, Maya cooking recipes, Maya ceremonies, Maya plant uses and many things that the Mayas practiced from way back in the rainforest. There are many books which are written about the Maya people and our traditions and wisdom, but this is a book written by myself, a Maya woman, who had the Maya ways passed down to me through many generations. You will learn in this book how the Ancient Mayas practiced women's wellness with womb massage, herbal steam baths, good nutrition, a happy life and herbs. You will learn how Maya men and women maintain their physical and spiritual health and that of their families with special prayers, ceremonies, herbal baths, medicinal teas and special care to their bodies and spirits. I have included all of this so that you can put it to practice in your daily life, if you so wish. It is my hope that these teachings will be preserved for many generations to come.

I hope you enjoy my book and that it will serve to clear your mind, and to help you find your own truth and faith and your own good way of living.

Miss Beatrice after a Primicia (Photo By Monica Juchum)

Chapter One: My Heritage and my Birth

Miss Beatrice's Grandmother, Delfina Tun

"My grandmother said that when I was born, I had the caul over my face all the way to my navel. In my culture, being born in the caul symbolizes that the baby is going to be a spiritual healer and a seer."

My name is Miss Beatrice Torres Waight. I am one hundred percent Yucateca Maya and Maya is my first language. I come from a long line of healers, midwives and shamans. My grandmother, on my mother's side, was Delfina Zib Tun. She was a midwife and an herbalist, as were her mother and grandmothers before her. My maternal grandfather died before my mother even knew him. His name was Guillermo Guillen (pronounced " gee yermo geeyen"). My maternal grandmother then married Pedro Tun who became my mother's true father. Pedro was a healer and a shaman who came from many generations of shamans. Everybody loved him as the true father of the family. He was truly my Grandfather.

My Paternal Grandmother was named Anastacia Euan de Torres (pronounced "aywan day Torehs"). She was a midwife and an herbalist and her husband, my father's father, was a snake doctor, a healer who specializes in treating snake bites.

My father was named Alejandro Torres. He was a Maya healer, a ceremony leader, a massage therapist, a shaman, a farmer, an herbalist, a snake doctor, a bone setter and an acupuncturist in his own way. My father used thorns called scorpion thorns from the female scorpion plant for acupuncture treatments instead of needles. Sometimes a snake tooth was also used. Mayas have their own system of acupuncture which is somewhat similar to Chinese acupuncture. Maya acupuncture is used to move wind out of the body in the case of wind invasion and to move stagnant blood. My father often used nine thorns in a cross form with prayers.

We Maya have many powerful healing practices. We also hold many mysteries. Amongst ourselves, we do not call ourselves Mayas, we call ourselves *"Masewal"* because our true origin is unknown.

Some people say we came from Tibet a long, long time ago and some say we emerged from a cave near México City even further back in time; but nobody knows for sure just where we Maya migrated from. Up until this day, it remains a mystery.

My mother, Dominga Guillen de Torres was a midwife, a mama, a nurse and also an economist because she took care of us very well. Both sets of my grandparents were born in Vallodolid in the Yucatan Peninsula of México near the city of Merida. For my grandparents, life in México was very difficult. There was famine because the land was dry and parched.

Also, the Caste War was going on. It was called the Caste War because Mayas were revolting against the caste system that the Spanish were enforcing upon them. The Mayas were also fighting for their homeland. The Caste War lasted from 1847-1901. During those years, many people escaped from México. This was very dangerous. The Mexican government, called the *Alcalde*, would shoot anyone who tried to leave México, calling them a traitor. It was one of the most violent conflicts of the Maya people. The *Alcalde* would do whatever they wanted. If they wanted someone's woman, they would kill her man and take her for themselves. If someone had many daughters, they would take some of them and make them into slaves. A group, including my mother's and father's parents, fled from Valladolid during the war to spare the men from being sent to war and the women from being separated from their children and forced to take care of the fighting soldiers. They migrated from the Yucatan to a Guatemalan settlement called *Chuch Kitam* which means "intestine of the wild pig." They walked through the jungle by day and on the road by night as not to get caught. Thank God, they made it.

They stayed in *Chuch Kitam,* which was right on the border of Belize, for about 10 years. In Chuch Kitam, my grandparents and their little group were constantly harassed by outlaws called banditos. To relieve themselves of this trouble, they migrated to another little town in Northern Mexico on the border of Belize called *Ik k'aay Che',* which means "wind is singing in the tree." We pronounce it, "eek ky chay." It seemed like nobody could bother them there because it was so close to Belize. Belize was a safe place with a better government. The British were ruling in Belize, which at that time was called British Honduras. My father was born in Ik k'aay Che' in 1905 and my mother was born in 1915.

The group stayed in Ik k'aay Che' for a long time. They became known as the Ik k'aay Che' Mayas. However, life in Ik k'aay Che' was not entirely peaceful. A group of people called the *Crusoob* were raiding them regularly. The *Crusoob* lived in a town called Santa Cruz, also known as *U Noh Kaah Balam Nah Chan,* in what is now the Mexican state of Quintana Roo near Ik k'aay Che'. *U Noh Kaah Balam Nah chan* means, "the place of the little jaguar house." The *Crusoob* were a mixed people descended from shipwrecked Spaniards, refugee *Arawak* people and *Carib* Islanders from the north coast of South America and the Southern West Indies. They were also mixed with escaped African slaves. The Crusoob were very strong fighters who had forced recruitments who fought alongside them. About every three months, they would raid Ik K'aay Che' and steal corn, rice, beans, chickens, pigs and horses.

Finally, my ancestors, the Ik K'aay Che Mayas, decided to put an end to the Santa Cruz raids. A group of Maya elders came together in counsel to make a plan. My father, Alejandro Torres, became the leader of the last war against the Crusoob. The Ik k'aay Che' Maya secretly spread a thorny plant called Cockspur on all the passageways out of the village and sneakily covered them with leaves so that they were invisible. They also made a homemade bomb that would go off when the Crusoob entered the village by the main road which was the only place where there were no thorny plants laid down. When the Crusoob entered Ik K'aay Che', the Ik K'aay Che' Maya were prepared to fight. They killed all of the 300 Crusoob who had raided the village that day. The Crusoob could not escape because when they tried to flee by the back roads, their feet were torn by the cockspur. The Mayas laid the Crusoob bodies in a big ditch that they dug in the ground. No Ik K'aay Che' Mayas died. It was a great victory. However, they soon began to fear that another army of Crusoob would form to take revenge on them, so they migrated into Belize to a place called San Jose, Yalbec. *Yalbec* means "Big deer have big antlers."

When my grandfather stepped into Belize, he said to the group, "We must never leave this place. It is a paradise on earth!"

My mother and father got married after they had landed in Belize. However, in Yalbec there came some more trouble. Yalbec was owned by Belize Estate Company (BEC). BEC owned most of the land in Belize. Initially, the manager of BEC, Mr. Brown, allowed the Ik K'aay Che' Maya to settle there, but then changed his mind a few years later. He decided that he wanted to start logging on the land and harvest lumber there, and the Maya were only in the way of his project. He came around with a paper and told everyone that if they signed the paper they could have lots and lots of land. Without even thinking, the people became greedy and thoughtless and signed the paper, even though they could not read it. In reality, the paper was an eviction notice, and thus they had unknowingly agreed to leave Yalbec. Once they realized they had been tricked, the Mayas refused to leave yet another home that they had settled into, so Mr. Brown ordered his workers to burn down all their homes. Needless to say, the Ik K'aay Che' Maya moved on once again.

The majority of the group headed to San Jose Palmar in the Orange walk area of Belize, but my Father said, "I don't want to do that! If most of the people are going there, let them go! I'm not going to follow them. No way am I going to that swampy place where there are mosquitoes biting all the time!" They heard that down here in the Cayo district of Belize, the mosquitoes are only seasonal, so a small group of seven families, including my parents and their parents, decided to head to Cayo district, to Santa Familia, where we are right now. The number seven is a sacred number to the Maya and finally the weary group found some good luck!

When the small group migrated here, they had to walk long hours and pass three days without food. It was 1927 and my mother was pregnant with her first son. They left all their corn, beans, chickens, pigs and everything behind, because how much can two mules carry? Besides, so much was lost in the fires. They couldn't bring much at all. It took them three days to move from Northern Belize to here. When they arrived here in Cayo, my grandfather, Pedro, said, "I think we have hit paradise. We are not going to move from here. The only place we are going to move from here is to the cemetery." Well, he ended up to be totally right!

When my family first came to Santa Familia, it was called the "Village of Johnny Sam." The village was named after its owner's older brothers, Johnny and Sam. Originally, Johnny, Sam, Arthur and James Humes owned the place through their hard work and dedication. James, the youngest brother, ultimately inherited it. It was Mr. James Humes who welcomed the Maya families to come and pick their piece of land for one hundred and fifty Belize dollars (seventy five United States dollars) per year. It was hard for them to get that amount of money, but this place was very good for them, and so somehow they came up with the money. It was well worth it.

The Mayas settled in Johnny Sam and began practicing their traditional ways of agriculture. They grew cassava, cacao, sweet potatoes, beans, corn and rice. They reared animals such as pigs, chickens and cows. They had finally found a place where they could safely live and practice their Maya way of life. The families had many children and the village expanded. Things were going so well that Mr. Humes changed the name of the village to "Bright Lookout" in the 1930's.

After some time, a Catholic priest Named Friar Rataman decided to build a school so that the children of the village could be educated. The whole village worked together to build the school which would also serve as the church and a house for the teacher. The families provided the priest with his basic necessities to show their gratitude. By 1950, the school was built. The priest suggested that they change the name of the town to Santa Familia, which means Holy Family, because of how well they all worked together and helped each other like one big family. Everyone agreed and the name was changed.

During the 1950's, when George Price was campaigning for an Independent Belize, the people of Santa Familia asked him to buy the land from Mr. Humes so they could be relieved of paying the high fee to live here. He bought the town and that is how Santa Familia came to be independent.

This place, Santa Familia, was a vibrant jungle full of everything we needed. In the wild, we could get the wild cohune (coconut), the wild grapes and wild apples. We could get wild papaya and little fruits called *Kayep Waya* which didn't totally fill you up, but if you drank some of the juice, it gave you lots of energy. There were sapodilly trees to make chewing gum, wild cherries and also mami apples. There was wild spinach and wild amaranth which we call *kalaloo* and wild allspice growing all around. There was ramon which is wild breadnut, growing from which we made our tortillas and porridge. Ramon leaves were also used to feed the animals. In breadnut season, you could get bags and bags and bags of it, so that's why my grandfather said, "This is a paradise."

Over there in México, it was always dry and stony. There was very little food; so when my Grandfather came here and saw that everything was so abundant, it was a paradise to him. There were also many fish in the river and abundant game meat in the jungle. There were lots of medicinal and edible roots in the forest. Back then, we could just go into the forest and bring deer, antelope, pigeons and pheasants home for our dinner.

There were Cohune palms. We made our roofs from the leaves, oil from the nuts and we ate hearts of palm as a delicacy. The outer shell of the palm nut was used for fuel for our fire heart. In those times, we had a fire heart which we called our *fogan.* We cooked with the charcoal from the wood we harvested. *A fire heart is a place in the ground like a little square filled with dirt or a little elevated place where we keep our fire. The fire heart is the center of our home, our culture and our spirituality.* I still have my fire heart today where we cook our corn for our *tamales* and *bols* in preparation for our ceremonies and special occasions.

In those times, nothing was ever wasted. We really appreciated what we had. If we had any leftover scraps from harvesting and processing our food, we used it for animal food. You name it and it was here! It was all here because there were not a lot of people and the jungle was still wild. This place was very good for my ancestors. At last they had found a home.

When my parents arrived in April 1927, they settled right here in Santa Familia right across from my home, where they lived all their lives. They had eight living children, including me, and they never moved from that land. I was born right on that land in the original house. It was made of little bricks and leaves. They were always fixing it and patching it until my father died. Then there was nobody left to patch it, so it receded into the jungle. Hurricane Hattie blew away a piece of it, but it is still there and it is still standing. My mother didn't want to tear it down because of all the good memories that stay there.

After a while, one of my brothers told her, "You cannot live there anymore! That house is too bad for you! How do you know one of these days a little wind will not come along and blow it right on top of you."

So finally she gave in and let him build her a comfortable house on the same land. There she lived for the rest of her life. It's next door. We can walk there from my home.

My mother had thirteen children, but only eight of them lived. Three of her children died in an epidemic of the measles and whooping cough. One little girl died of the evil eye. The other one ate food that was too hot and then got a chill; and it clouded his blood and he died of *Pasmo*. *Pasmo* is a disease in Maya medicine which is caused by being exposed to a cold draft, cold water or cold food and drink when the body is overheated.

I was born in Santa Familia in the Cayo district of Belize, Central America on the 18th of January 1948 at approximately nine o' clock in the evening. The day I was born was a Sunday. I am the 7th child. We Maya believe that seven is a sacred number and Sunday's child is full of Grace. Children born on Sundays are sensitive and worshipping. They honor the saints and usually become a nun, a priest or a spiritual person. I was never meant to be a wife and a mother. I was meant to be a spiritual person like a nun or something. Nobody ever pushed me in that direction, so I didn't go there, but I am a spiritual person none the less.

My grandmother had a fever on the day I was born, so she couldn't deliver me. My grandmother was in the process of training a new midwife, by the name of Domitila Rivera. Domitila was the only one available to attend my birth. My mother said this lady didn't know much because she didn't give her the tea or massage her or warm her belly properly like my grandmother would have done. All the little things were missing.

"Maybe she was learning midwifery properly but she just hadn't really gotten it yet." My mother proclaimed, giving her the benefit of the doubt. "With how fast Beatrice came, I didn't really *need* any special labor care because Beatrice was so ready to come and she was coming fast!"

My grandmother said that Domitila told her that when I was born, I had the caul over my face all the way to my navel. In my culture, being born in the caul symbolizes that the baby is going to be a spiritual healer and a seer. We call the caul, *manto*. It is the amniotic sac. Mine was a big one. Normally the caul is only over the face and neck but mine went all the way to my belly button. The midwife had to rip it really quick or else I would choke to death because I was not breathing! Immediately when that was done, I was crying and crying because I was hungry and to my mother's surprise, when she put me to the breast, she noticed I was piercing it with a tooth! She panicked because that had never happened to her before. She knew it was a sign but she didn't know what it meant. She kept it a secret. She didn't even tell my father or my grandmother. On the third day of my life, the tooth disappeared. My mother panicked even more because we believe if you swallow a tooth, you could die. She fretted and she worried and she looked all around but she couldn't find that tooth. She said that the tooth either disappeared or I must have swallowed it. Up to this day it, it is still a mystery.

Maya Plant Ally Number One: Siempre Viva

Kalanchoe Pinnata

Life everlasting plant in Miss Beatrice's back yard

(Photo by Katherine Silva)

Life everlasting is a good plant ally to have. It is also called Tree of Life, or in Spanish it is called *Siempre Viva*. Here in Belize, we use this plant to make a bath that cures achy joints caused by wind invasion. Siempre Viva gives you a little hint into its uses. This is the doctrine of signatures. It has little joints on it which tells you that it is good for achy joints. In Maya healing, we make a poultice with Siempre Viva plant by mashing the leaves and applying them to the area of the body needing healing. We apply the poultice to the head for headaches or to draw out a fever, to the sacrum for inflammation of the tailbone or sacrum and to the chest for asthma. Also for asthma we squeeze the juice of three leaves into a spoonful of raw honey and take it internally for three days. For diabetes, we eat three leaves of this special plant at every meal. Don't try to make an herbal oil with this plant because it is far too watery! Siempre Viva grows all around my house and if you take one leaf and put it on the earth, you will have another Siempre Viva plant in quick time!

Maya Altar in Miss Beatrice's home (photo by Katherine Silva)

"The egg is included in the ceremonial tortilla to symbolize opening the baby's mind. The corn is included because it is a sacred food for the Maya. The chaya is there because the baby will feed on that to nourish his body for his whole life and the pumpkin seed, the siquil, is there representing a new little sprout. The baby to be blessed is a new little sprout who is going to germinate and become fruitful."

When I was a child, many of the Maya ceremonies were done in my village. The *Hetz Mec* baby blessing ceremony, the menarch ceremony, the *primicia* ceremonies and many others were done regularly. Our ceremonies marked the special events in our lives and the cycles of nature and farming. They also held teachings within them that showed us a good way of living. My father, Alejandro Torres, was the spiritual leader in my village, as well as being a farmer, a spiritual healer, a bone setter, an herbalist, a massage therapist and a Maya acupuncturist. When he passed away, many sacred teachings were lost to us. Nowadays, the people in my village have forgotten about the ceremonies. I am the only one left here, keeping them alive.

When I was eight months old, I received the *Hetz Mec* ceremony. *Hetz Mec* means "an open embrace." *Hetz Mec* is a Maya ceremony practiced from way back. It is offered to babies when they turn eight months old, before they take their first footstep. Hetz Mec represents the initiation of the baby's first footstep into the world of prosperity and hope. The ceremony is always done on a full moon at dusk.

Preparations start well before the day of the ceremony. The Hetz Mec ceremony is a family gathering, and only the closest relatives like mother, father, grandparents and great grandparents partake. When the parent talks to the family members about doing this ceremony, he or she has to show very much respect, because attending the ceremony is a great responsibility. The invited persons take time to think it over. They will come by the house where the baby lives within two weeks time and let the parents know the final answer about coming to the ceremony. The father of the baby is responsible for choosing the Godparents. If the Godmother agrees to be the Godmother and attend the ceremony, she must provide a little white suit for the baby to wear at the ceremony and give two big pieces of white cloth to the baby, which symbolize the baby's wings. Every little thing that's done for the ceremony is meaningful. The ceremony is done on the full moon or three days before or three days after the full moon. When the moon is full, everything is charged with energy.

On the day of the ceremony, the preparations begin early in the morning. The mother of the baby prepares the food and the father of the baby prepares the place where the ceremony will occur. The preparations for the ceremony are very involved. First, we make a little altar. The altar is a round table with a white table cloth. A white candle is placed there on the altar to symbolize enlightenment. The white candle represents a clear way of guiding the child on the right steps to prosper in the future and gives the child illumination to continue the good practices of the Maya.

White Copal is burned under the table in an *incensario.* An *incensario* is a fancy clay pot with a handle in which we burn *copal,* our sacred incense, on a piece of charcoal. A copal burner in Maya is called *pul. Pul* means smoking. *Copal* is harvested in the jungle on the full moon with a ceremony and special prayers. *Copal* symbolizes the blood of the nine benevolent Maya spirits which is sacredness. White Copal is used when we are invoking spirits and asking for blessings.

A bouquet of white flowers and two pieces of white cloth are placed on the altar table. The two white pieces of cloth symbolize wings for the baby, so in case the baby dies the wings can carry the baby up to heaven, just like a little angel.

Everything in the ceremony is white. White symbolizes purity, a new beginning and sacredness. The baby is also dressed in white.

Special food is placed on the altar. Chocolate is served on the table because chocolate is an appropriate food for the Maya Gods. Little tortillas are also placed there. The tortillas are made with corn, eggs, *chaya* and *siquil. Chaya* is a nourishing green plant like spinach. *Siquil* is ground pumpkin seed. The egg is included in the ceremonial tortilla to symbolize opening the baby's mind. The corn is included because it is a sacred food for the Maya. The chaya is there because the baby will feed on that to nourish his body for his whole life and the siquil is like a sprout, just as the baby to be blessed is a new little sprout who is going to germinate and become fruitful.

The tortilla that we make for the Hetz Mec ceremony has to be really soft for the baby. It is baked on the *Comal* which is a round piece of clay or metal that is warmed in the fire heart. The *comal* is like a griddle for making tortillas.

A godmother and a godfather attend the ceremony to show their commitment to care for the child in case anything happens to his parents.

When the ceremony is about to start, an elder from the village says a Maya prayer and burns white copal in the incensario under the altar table.

When the ceremony begins, the godmother takes the baby on her right hip and goes around the right side of the table nine times. The baby is held on the right hip because the ceremony is done the Maya way and for Mayas going to the right, clockwise, is sacred and this means the baby's life will come out "right."

The godmother is the one who goes around the table first. This symbolizes that the godmother will be like a mother to the baby in case something tragic happens to his birth mother. Each time she goes around the altar, the godmother calls the baby's name. While she is going around the altar, she is dipping the tortilla in the chocolate and giving little bites to the baby. She is also asking God to bless the child and help him grow and become a well respected person, who respects others and is a productive member of the community. The godmother and the baby walk nine circles around a round table to honor the nine benevolent Maya spirits.

Why do we use a round table? Round is, think about it, round is like an unlimited friendship, never ending.

The round table represents unlimited friendship between the child and the godparents and the child and the world. For the Maya, it is very important to create an atmosphere of peace and friendship for our children because friendship helps us prosper in life.

To the Maya, a circle means friendship. We don't make a ceremonial circle with just anybody. For us, it is very important to go into circles only with people we know. Making a ceremonial circle is like a little pact that you are making with your friends that says, "You are there for me when I need you and I am there for you when you need me." In the Hetz Mec ceremony we are going around the circle to confirm our friendship. The godmother takes the baby around nine times to honor the nine benevolent Maya spirits. The godfather also goes around with the baby nine times and then turns the baby's little face to the south. He is faced to the south, the direction of abundance, of seeds, of fruits and flowers, to receive good energy and positive vibes. This helps the baby to be a person of good action.

On the last time around the table, the godparents promise that if anything should ever happen to the parents, they will care for the baby.

The parents of the baby provide festive food and drinks to those who came to attend the ceremony and together they all enjoy the feast.

Each year after the ceremony, the child is expected to offer something back to the community such as helping an elderly person or helping to bring a meal to a neighbor in need. Throughout the baby's life, not every day, just on special occasions, the godparents give small gifts to the baby, such as a little hammock or hat.

If the baby is a girl, the godmother teaches the child how to cook and sew, needle point, cross stitch and make thread out of cotton. For making thread we had a little stick and tons of cotton with a gourd bowl. We pulled and pulled the cotton until we had a big ball of yarn. With our home made yarn we made our own clothes and many useful things. For a boy, the godfather taught him the skill of hunting or tanning the leather and farming so that he could do something good in life and not be a lazy boy. Back then, they didn't go to school but they knew many skills.

The Hetz Mec ceremony was my first ceremony, but not my last. When my first menses came, there was another special Maya ceremony offered to me. When it was my time, I was taught many things and given a little ceremony. My mother used the same table for my menarch ceremony as she used for my Hetz Mec ceremony. It was a round wooden table with three holes in it. She put three little sticks in the holes for the legs. For babies and young women, we use a round table to signify everlasting friendship and the cycles of life.

In the Maya ceremony for first menses, my mother offered chocolate on the altar in a special blend and lit white candles. She burned white copal and offered special prayers.

The menarche ceremony is a special day for a young woman. All the grandmothers, great grandmothers, mothers, aunties and older sisters gather to share their wisdom with the maiden girl.

The elders teach the girl that her menses is a time of great power for her and that she needs to honor her sacred time. They teach the young woman to rest, dream and pray during her menses and not to give away her power. They tell the girl about how if she has sex now that she is menstruating, she can expect to get pregnant and they also teach her about all the things she needs to do to take care of herself.

The grandmothers taught us girls how to be careful, because way back then, nobody was teaching about contraceptives and nobody went and had sex before getting married. It was a rule in our culture not to have sex before marriage. So you see, it was very important for us to know what the consequences of sex could be and how to protect ourselves!

In the ceremony, the elders each speak to the girl. It's up to the girl if she wants to listen or not. If she listens, she'll learn ways to be okay in life and in society. And that's good.

If she doesn't want to listen, she may do unwise things and have to pay the consequences. This happens easily because when we start in a relationship with a man everything is good at first and then little by little it can go bad and then we could be pregnant with no father for the baby and no family to turn to and who can we call on then? One in a thousand men in that situation will take responsibility for his child and woman. Why get into this kind of mess?

Way back, you couldn't even visit your family if you went and got pregnant without being married. They would ignore you and disown you totally because you didn't honor them. Way back, if a girl got pregnant outside of marriage, she would be outcast. She would lose her family.

My father had even said, "Way back, if a girl had a baby before she was married, everything the people gave to her would have a hole in it. Even all the gourd bowls given to her had a hole."

I said to my father, "I think that is not fair. When someone gets pregnant, it's not just about one person, it's about two people. Why condemn the woman? Any consequence should apply to both people who participated. I don't think that is fair! I think that is too harsh on a woman!"

Women used to be belittled by men and that's not right. We want to do good things also. Thank God, little by little, more power is being given to women and I think that's really good. Because when people respect you and you respect yourself, you feel much better. You feel empowered and you can do much better. You can teach your kids that way too.

Nowadays, we are more accepting of our children's mistakes. Times have changed quite a bit. Now we believe that it's not good to turn out kids. We all make mistakes. We cannot be so perfect in life. Some might listen and some might not. I always say, get ready for the unexpected because we never know what will happen next. People may think, "*My* kid will never do that!" But, you never know, she may end up doing worse! We just never know what our children's path will be so we have to try and love them anyway!

After the teachings and ceremony, the mother handed me a little bag she had sewn herself. When I opened the bag I saw many pieces of cotton cloth inside.

My mother spoke to me, "This is your little baggie. From now on it is for you to use to catch your blood and to wash every month. The blood you wash out needs to be given to the earth or to a special plant and when you see how the plant is flowering and bearing fruits, you can be proud of yourself because you are taking care of that special plant. If it is a flower that you choose to put your blood on, oh boy, when it blooms, it's showing you how you yourself are blooming."

At the end of the ceremony, my mother burned white copal and said the closing prayers. She then cut up the bread and passed around the chocolate and shared it with the circle of women gathered there. I got that ceremony! I was lucky!

The traditional Maya marriage ceremonies had no priest and no pastor and they lasted for nine days. A *primicia* ceremony was held on the first day. The young bride didn't just go love the guy. She had to wait until the ninth day of the ceremony when there was a big feast and then she would be married and then she could go with her man.

Before the ceremony, her mama had to train her to do the washing by hand, clean the house, make the tortillas, grind the corn, cook the beans, dress the chickens and feed the chicks. Another part of the job of being a wife was to patch holes in pants and shirts. All those things a woman had to learn before she got married. If she didn't know how to do those things then she couldn't get married. It's as simple as that.

When the guy came and asked for her, her parents would say no, because she was not yet ready. If she didn't know how to tend the house yet, how could she be a wife? There were many other things a woman had to learn how to do, like how to get her own firewood, harvest her beans and corn, pack water for washing from the river and kindle her own fire. A woman needed to know how to make her own clothes sewn by hand and how to make little baby clothes before she could get married. Being married is a big task and not an easy one.

If you are thinking, "Oh, I want to leave my parents and go be happy with that guy," think again! It's not going to happen like that because marriage is a lot of hard work. I know from my own experience. Imagine, we even had to make our own oil! We collected our small wild Cohune or our big coconut. We would break them with a machete and use the shell for fire wood. With a mortar we would pound the coconut flesh really well and then boil it and collect the oil on top. We would strain it and that was our oil for cooking. Making oil is not an easy job. It is really hard work. Also, do not forget that we had to learn how to make tamales, bols, hominy, tortillas, salsa and traditional soup. There were no grinding machines. Grinding was done on a stone! I got up at three in the morning to get my tamales started and done by five. Making tamales is a lot of work. If there wasn't much fire wood, I had to blow it and blow it to keep it going. It's a really hard job. It's even harder if the wood is wet or it's not the right kind of wood. The other task that I did as a married woman was to gather the corn when it was ripe and select the bigger ears of corn to stick in a square container, which we called a *troja*. It was a secure place to store our corn. I even had to pour ashes on top of the *troja* so that the termites wouldn't get it. Then the corn lasted the whole next year.

Women work harder than men. The task is never ending! Wash the dishes, cook the food, take care of the babies and do the washing.

It's a hard sacrifice. When it was mushroom time, we would collect small little mushrooms that we went really far to get. Mayas like to cook fresh food three times a day instead of eating leftovers so that is a lot of work. Being married does have some rewards, but I tell my daughters to think it over really well before they get married, because it is a big job!

For menopause, we also have a special Maya ceremony. We give thanks to the universe for getting us to that stage because many women don't make it to that stage. We give thanks and light a little candle and offer a ceremony with corn porridge and white flowers. All offerings are welcome to give thanks and gratitude. We remember that we have been fertile and fruitful and now we are stepping onto the next path. We give thanks for all we have achieved in life so far and then we burn copal of amber color. Amber is the color of copal that is showing gratitude to the benevolent Maya spirits.

I encourage women to make a gratitude ceremony when they reach menopause. You also can do it! If you don't have amber copal, you could burn white sage, sweet grass, artemesia or dried basil as an offering of gratitude to whosoever is your deity. Do not burn marigold in your ceremony. Marigold is to ward off evil and this is a thanksgiving ceremony. You can do this in your own way with your own deities and your own plant allies. What is meaningful to you is what matters.

Before the ceremony, you could take *un Bajo Sagrado*, a plant spirit bath for a *limpia*, a cleansing. You could collect your plants: basil, marigold, rosemary and red roses and make your bath to take before you enter into the menopause ceremony or circle. I hope you will enjoy your ceremony as much as I did mine. It is an important way to honor yourself and your womanhood.

Along with the ceremonies for each stage of life, the people in my village honored the seasons and the cycles of planting and harvesting in the *Milpa*. My father held agricultural ceremonies called *primicias* many times each year. The *primicia* is the first mass that the Mayas offered to ask permission to till the land, to ask for blessings on the seeds and to give thanks to the beneficial Maya spirits for a good harvest. *Primicia* actually means "first fruits." My father always held a primicia to bless the seeds and help them germinate and to give thanks for the first crop. As children, we were told that we should never touch that first crop until it had been blessed with the primicia.

After the corn harvest, my father would cook a whole big pot of corn porridge sweetened with wild honey and offer it to the nine benevolent Maya spirits during the ceremony. After the primicia, he would share the corn porridge with the neighbors, his close friends, and still there would be lots left over for us! At the primicia, he would burn his special white copal. He would brush off each person entering the ceremony, with special healing herbs, such as wild basil, john charles, blue vervain or neroli. He would use nine sprigs of one of his special plants to do the plant brushing.

For primicia ceremonies, he used a square table because he needed four corners to represent the four corners of the earth.

He would place nine gourd bowls of corn porridge on the altar as an offering to each one of the nine beneficial Maya spirits. When people came to the ceremony, they were expected to partake of the feast of corn porridge, to take at least a little sip. If they didn't want to partake, it was best for them not to enter, because refusing to taste the corn porridge offends the spirits in the ceremony.

All of our generations grew up eating corn tortillas three times a day and drinking warm corn porridge every day. Our corn porridge is called *sa'* or *atole*, which is made with corn, water, wild honey or *panela*. *Panela* is like brown sugar. It is the whole brown sugar straight from the cane. It has a salty taste. *Panela* has to be carefully selected and harvested at just the right time and processed properly.

We grew our own corn in a sacred way, but when we do this ceremony now sometimes we buy *masa harina*. That bag of masa harina is only used for primicia ceremonies, not for breakfast or ordinary cooking. This is also true for the honey we use to sweeten the porridge because it is sacred and for one purpose only. The gourd bowls as well are only to be used for the primicias. In this way, we keep everything to be offered pure and beautiful. When we make the corn porridge, if there is some left over after the ceremony and feast, we give it back to the earth. It is even better to drink up every last bit of porridge. The ceremony actually is not over until all the porridge is gone. In our village, we use corn porridge for primicias. Chocolate is never offered. Chocolate is offered at the Hetz Mec baby blessing ceremony, the menarche ceremony and the menopause ceremony, but not at primicias.

I have found a way of keeping the primicia ceremonies alive for my family and for my students and clients who have faith in the Maya ways. We offer primicias on the full moon. We take a purifying herbal bath before the ceremony and we dress in white. On the altar table we place a clean white tablecloth which is only used for primicias. On the altar we place nine white roses, two angels, a crystal glass of water and four white candles for the four corners of the earth. We also place something we harvested from our garden, if we have one, like some ripe tomatoes, a pumpkin or some flowers. If we have something from our first harvest that is the best. That way we can offer thanks for the whole crop before we enjoy it.

When the ceremony begins, we offer the specific primicia prayers to the four corners of the earth and lastly to the earth itself. We chant two times in each corner and then one time to the earth. At this point we create a quiet space for the spirits to enjoy our offerings. In the silence, we meditate upon our gratitude for a great harvest or the blessings in our lives. If there is something we really need, not something we simply want, we focus on that as well. We focus strongly on our need during the meditation. It can be a need for blessings on our seeds, a financial need, needing a job, a healing need, desiring a baby, needing a husband or anything we really need for our well being. It is best, however, that we choose only one need to focus on. That makes it more potent, more powerful! For example, you could focus on, "I need a job." Whatever we focus on will come true, usually within one month's time. Remember, we must only pick one thing. We want the benevolent spirits to fully focus on our one big need and not to be spread thin taking care of many little requests. In this way, we keep our minds clear and open to receive. For all the other people that we want to pray for, we write prayers on little pieces of white paper to be put on the altar. These are requests for help for other people such as "for my mother's health," "for my kid's happiness" or other similar requests. We can put many prayers for many people on papers on the altar and the spirits will attend to those, but for ourselves we just internally focus on one thing.

For example, you may need a dream vision because you know you have to move to a new home and you need some guidance. If you ask for a dream vision at the primicia, you will get that vision for sure. The benevolent spirits come to enjoy the feast, to receive our gratitude and to help us.

When the meditation is over, we ring a little bell to tell everyone to come back. A final set of prayers is whispered to the earth. To complete the ceremony we all sip the corn porridge.

When the primicia is over, the candles are not blown out with our breath. We spit on our thumb and finger to put the flame out or use a candle snuffer. We never blow out a candle that has been used for a sacred ceremony because then we disrespect the spirit of the fire and blow away the good luck!

When I was a child, children were not allowed in the primicia ceremonies because they are sometimes noisy, and it is harder for them to keep quiet and concentrate. My father was concerned that their behavior might insult the spirits. As we grew older, we were accepted into the primicia. Way back, we couldn't attend a primicia until we were thirty years old because it was a very serious ceremony. At that time, a small group who were all healers, in their own way, came to the primicias. The people coming to the primicias in my village were a little group that believed in ceremonies. Mostly they were part of the family: the great uncles, the grandfathers, the really close friends, (the *compadres*). They were a little family that stuck together and they did their prayers together. They were all healers and they had to follow a straight path in order to attend these powerful ceremonies. For example, if there was a man who was cheating on his wife, he was not allowed in.

Also, women who were bleeding were not accepted into the primicia either. This was because bleeding women are very much honored in Maya culture. They are having their own ceremony. Women are also very open during their menses. It's possible that something negative being released during the ceremony could enter them. This guideline was there to protect women.

Nowadays anyone who can behave can be in my ceremonies because I believe that everybody needs a blessing. I welcome everyone into my primicias. There is one exception, however, that I like to abide by. When women are menstruating, it is best that they not come into this particular ceremony. It is not because they are not welcome there. It is actually to protect them. When women are bleeding with their monthly cycle, they are very open, very spiritual and very vulnerable. Sometimes in a primicia, people are letting go of dark things that they no longer need or healing a deep pain or a severe trauma, and a bleeding woman could be negatively affected by those energies. A woman's moontime is a special time for her to nurture herself and connect with her own guidance and so she needs to create a sacred, clear space for herself and have her own ceremony. She needs to be honored during her sacred time and be given her own space to rest, dream and be pampered.

I am the only one left in my village doing primicia ceremonies and I have lead them all over the world for women, men and children. To participate in a full primicia, you need to find someone like myself who is trained properly and initiated into leading a primicia ceremony and who knows the proper prayers and the correct way to offer this ceremony.

When I was young, the primicia was only lead by my father. Children weren't allowed in and it wasn't considered to be something for a woman to lead. Because of that, I was trained only a little by my father while he was living, but I was watching how he prepared for the ceremony and how he handled the porridge, the herbs and all the little things that had to be done. He wanted me to be a midwife, not a ceremony leader, but none of his sons took up the task of following in his footsteps, so I was trained and initiated by him in the dream time after he passed away. I was the only one who was ready to take the responsibility.

I have also learned how other people offer their primicias and I have revived my family's rich tradition through my hard work and dedication and am committed to keeping the ceremony alive and well. I have trained and initiated a few women in the US in the leading of primicias, so that the ceremony will continue. Very sacred powerful ways can be totally lost if from generation to generation they are not passed down. If there is a break in that chain even just once, entire treasures can be lost. I have done my best to pick up the thread of this line of healers extending way back and to walk with it into future generations.

The full primicia is no joke. It is very powerful because it has been passed down through many generations. It stretches back to the ancient ones and must be approached with great respect. It must only be conducted by someone who has been trained and initiated properly. Be careful, because if you try it without training and permission, you could get something you didn't buy. I am putting the information about primicias in my book so that when you attend one lead by someone who is trained and initiated to conduct it, you will understand a little bit more about this special ceremony and the rich Maya culture. You will know how to attend this ceremony respectfully and reap the benefits.

Also primicias should not be photographed or recorded during the ceremony. You could take a picture before or after of the altar and the friends gathered there but never during the actual ceremony.

I trust that my readers will respect the sacredness and power of these things I am telling you.

But don't worry. Anyone who has faith can make their own mini primicia at home, on the full moon, with a white cloth, a candle, a glass of water, a flower and your own special prayers. This will bring many blessings into your heart and into your home.

During my childhood, my family practiced both our Maya way of life and Catholicism. There was no conflict between the two paths. They complimented each other nicely.

When my father was living and it was Easter time, he would call everyone together and tell us what we would do to celebrate this special time. He was the spiritual leader in my family and of our little group of friends and relatives. He always made a special ceremony for Easter. He gave us a blessing every Friday of the holy season which ends with Holy Friday. The first day of Lent is Ash Wednesday and then there are seven special Fridays that follow. For the seven Fridays, my father would make a little porridge with a special kind of corn which he called *sak kaab.*

We don't cook *sak kaab* in the ordinary way. We just take the ear of the corn, shell it, grind it on the grinding stone or in the grinding machine, boil it until it gets soft and then sweeten it with wild virgin honey. No lime is used to make *sak kaab.* The corn is first boiled until soft and then ground. It is a watery porridge. *Sak kaab* means "white honey from virgin bees." These are tiny bees that live in holes in the sides of houses and they make the best honey for medicine and ceremony. On the Thursdays, he offered sak kaab to the *Bokan* of the earth by putting it in little gourd bowls, very small calabashes, which he would put all around the house strung on a holder he made out of little vines.

On Friday mornings during the holy season, before we would have breakfast, he would call everyone by age and give them a special blessing with the sak kaab. With his finger, he made seven crosses on our fronts, our wrists, our elbows, our foreheads, our chests and on our stomachs, while he whispered the Maya prayers. The blessing protected us from accidents and all kinds of sickness. It worked because we didn't get sick or have falls. Sometimes we would get a little tummy ache but that wasn't serious. My mother gave us raw garlic to eat and that took care of it. On all those Fridays, we were not allowed to eat any meat. We ate only vegetables, mostly hearts of palm and fish, but no chicken. On Good Friday, a big turkey that my mother raised was killed as a special treat, and we made a special lunch to be shared with the whole extended family. It was cooked in a big pot because a big turkey gives a lot of soup. This soup is called *chirmole,* which is our black soup, yup, that's it! And it was very, very good! Our culture is rich and beautiful and we use ceremonies to honor all the crossroads and special holidays in our lives.

Maya Plant Ally Number Two: Hibiscus

Hibiscus Rosa-Sinensis

red hibiscus flower, Placencia Belize (photo by Katherine Silva)

One of my all time favorites, hibiscus flower, which we Belizeans call *Tulipán*, grows all around my house. I use hibiscus tea to slow bleeding and to build blood. For hemorrhage, I put four open blossoms and five closed blossoms with nine leaves in one pint of water. I boil it for ten minutes, strain, and let it cool. I offer it to the lady to sip cool until it's gone. The tea must be cool because cool contracts and helps the uterus to close up and stop the bleeding. Warm and hot expands, so that may make matters worse by opening up the flow. I never make the tea icy cold because too cold is not good. We Maya never put cold into our bodies.

Hibiscus is also a wonderful blood building tea for women at any point in life. Especially in pregnancy, hibiscus tea provides women with the nutrition that they need. For everyone it is a nutritious, hydrating drink.

Hibiscus contains mucilage, which is something we lack when we head into menopause. That is why we need to eat hibiscus flowers in a green salad three times a day. It takes care of dry vaginas and stiff achy joints.

Hibiscus leaves and flowers are also good for your herbal baths and to make a soothing bath for skin conditions. A bath made of hibiscus is soothing and cooling. I drink hibiscus tea every day because it is full of iron and good minerals. It tastes so good, like a nice red punch. It is sooooooooo refreshing!

In Belize, we have a legend about the birth of Hibiscus. A long time ago there was a beautiful young maiden who was indigenous to Belize who lived near a great river. Her name was Tulipán. One day when she went to the river to fetch water for her family she met a young African man and fell in love with him. Every day, they met while fetching water from the river and so their love affair began. Each day Tulipán's trips to the river were lasting longer and longer. In those days there was a rule that you should not marry someone from another culture. Tulipán didn't care what color her lover was. She was very much in love with his soul. After a while Tulipán's father became suspicious about her river trips and why they went on for many hours, so he began to ask the villagers what she was doing by the river. The nosy people told him all about her affair with the young African man. Her father became furious and warned her to stay away from him. The parents of the African man had also warned him not to break the cultural rules. The young lovers' attachment to one another was so deep that they couldn't listen to their parents' advice. They often sneakily met by the river after dark. The parents' anger increased and rumors in the villages were very thick so the two lovers decided to drown themselves in the river so that they could always be together without anyone harassing them. Their bodies died in the river, but their souls rose up out of the water and transformed. Tulipán became the beautiful hibiscus flower and the young man became the hummingbird. To this day, wherever the beautiful red hibiscus flower is blooming and dancing in the breeze, there hummingbird will be also, chasing her and drinking in her delicious nectar.

Chapter Three: How I Found my Healing Path

Miss Beatrice's Healing Hut, Santa Familia, Belize (Photo By Katherine Silva)

"I had a dream vision of my father coming to tell me that I needed to take his place. In the dream, I said, "How in the world am I going to do that?"

When I was a little girl, I could see visions, fairies and spirits. I could see many things that nobody else could see. The spiritual world is very powerful and there were times that I was scared to death because of what I could see and experience. If I slept on the side of the bed, I would be awakened by an invisible "somebody" pulling me to the ground. Because of those spirits bothering me, I never could sleep on the side of the bed. I always had to sleep in the middle where the spirits didn't touch me. That was where I felt safe. Up until this day, I cannot sleep in the dark. I have to have a little bit of light. The dark is not a place where I want to be. I also light my candles and say my prayers so I can sleep with the angels.

When I got a little bit older, I think my mother told my father about what I was experiencing and how scared I was, especially in the dark. That's when he started doing special blessings and herbal baths for me. He began burning black copal in my room. My father also taught me how to do the "let go" ritual. He showed me how to put all my worries into nine little stones and throw them over my shoulder into the river. That helped me a lot.

When I was a child, we had all the food we could want. My father always lived by farming. In every part of his life, my father really practiced what he had learned from his Maya ancestors. My father had his *milpa* (a plot of land for farming) up on the hill where he grew tons of cherry tomatoes, bell peppers, *jalapeños, habaneros,* big white peppers, bird peppers, chives, tomatoes and pumpkins of all different shapes. We also had *jicama,* sweet potatoes, regular yams, purple yams, china yams, tarot root and beans. He grew black beans, lima beans, white beans, pink black beans and black white beans. We never suffered from a lack of beans. We always had plenty. We children had to collect them and shell them but we didn't mind at all. We liked to eat them. We also picked tomatoes, peppers, *jícama* and lots of other delicious home grown food. We walked all the way up to the milpa, three miles away, to bring home our food. We would bring home tons of peppers, chives and onions in big bags. We never had to buy our vegetables. My father watered his milpa with a little creek. His garden was flat and naturally damp so he didn't have to do much watering. The milpa that my father tended consisted of two acres of land.

Every seventh year he left a patch to rest. He taught us, "Every seven years, you have to leave your land to rest." He always respected the earth by farming in the traditional Maya way.

My father also grew plantains, bananas and sugar cane, but when my grandfather passed away, he stopped growing sugar cane because he needed his father's help to process it. Making sugar is a big job and the cane has to be cut at just the right time. If you cut the cane when it is older, the syrup will never harden. My father and grandfather made their own sugar cane blocks. I can show you; I have some pieces. They made their own sugar in a big hole. They would put the sugar cane in the hole and smash it with a stick. There was a container to catch the juices coming out. They would slow cook the juice on the fire for one or two days until it thickened, and then they would put it in little molds made of flattened cohune palms to harden. Sugar is a lot of work and very particular, so my father didn't want to do it by himself. After my grandfather died, we had to buy our sugar at the store.

My father was a bone setter, a snake doctor, an herbalist, an acupuncturist and he lead ceremonies. He was a busy guy! He worked for nobody but himself and we never went to work or school on an empty stomach.

If my father had a patient to tend to, he would stay home and work on the patient for half a day, and the other half of the day, he would work on projects at home. He didn't go to the farm on days when people came for healing because the midday sun in Belize is too hot for anyone to be hiking three miles up to the milpa.

Once, when my father was a child, he almost got killed by a horse while he was out collecting firewood. When he saddled and mounted the horse, the horse got startled and started kicking and threw him off. Thank God he fell close to a big tree stump and not right on top of it. It was a big fall and for days he couldn't walk, and from that day onward he said goodbye to horses! He often said, "I had better walk, even if I have to take two trips." Since that fall, my father never rode a horse; he just stayed on his own two feet.

When I was a little girl, I had many hardships. I had to lug water from the river. The neighbors would say, "I want ten gallons of water!" They would only give me five pennies for it. They would call out to me, "I need you to grind my *masa* for me."

It was a lot of work! There was an old lady living nearby and she was not very nice to me. She and her husband were elderly and she would ask my mother if I could come and stay with them when her husband went to México for an errand.

My mother would agree, "Yes, she will come and help you."

So for the whole week or two that her husband was away, I had to help her pick the eggs, fetch the water, get the fire wood, sweep the house and tend to the animals early in the morning. The old woman would fry a plantain and give me two little tiny pieces of it with an egg and a cup of coffee for breakfast. Her coffee was not even real coffee; she boiled it and boiled it and it was just like sugar water.

She would give it to me boiling hot shouting, "You have five minutes to have your breakfast!"

She would hurry me because I needed to run home, take a bath and get ready for school. I was always late. When the bell was ringing, I was running to get in line. At recess, I had to go to her house, wash up the kitchen, sweep the floor and grind the masa. There was no recess for me.

I really didn't have a childhood that I enjoyed. My childhood was all work, work, work. However, up until this day, I remember the good lesson that I got from that experience. I learned how to be strong and just go on in life. In the house and in life, there are always many things to do. You can never say that you are done. I do my best to keep a happy attitude and be grateful for what I have. Hard work doesn't bother me.

Because I didn't know Spanish or English, only Maya, I had a really hard time at school. Our teachers were *Garafunas* from Stan Creek and only spoke their own language and English. The *Garafunas* are a mixture of *Arawak, Carib* and West African people who travelled to Belize from Honduras in 1941. The language they speak is a combination of Carib and Arawak, with English, French and Spanish mixed in. It is very unique and really hard to understand. In school we were not allowed to speak Spanish or Maya, only English. I didn't know a word of English so, for me, it was like living in hell. Every day I was tormented.

In our family, my older sister Patricia and myself and two older brothers, Salustiano and Primitivo, were the only ones who learned Maya. The other ones did not learn our native language and I believe they missed something. They can understand Maya but they cannot speak it. Slowly, the use of our Maya language is fading away in our village. I still speak Maya with my mother, my auntie, my older sister and my older brother. Little by little, our language is getting lost, so I write Maya words in a little book for my children, with the English translation, so that they can learn a little of their true language.

When I was done with primary school, I worked for the richest family in Santa Familia Village, the Silvas. I worked for them doing washing. They paid me seventy five cents Belize for each big bundle of washing. I did that for three or four years and then I got tired of it. I wanted to study more.

I did have the opportunity to go to high school, but my parents wouldn't allow it. In Belize, you have to pay to go to higher education. I earned a scholarship from writing an essay. My essay won the contest and I got the scholarship. I was even going to get my books for free, my registration for free, plus a stipend. The only thing my parents had to do was buy me my uniform and my little tennis shoes.

"No way!" My parents scolded, "Education is not for girls; it is only for boys!" They didn't think that I needed to go.

I really wanted to go, but I didn't have anyone to buy me my clothes. The school was also very far away in Belize City.

My father continued, "No way! That is too far away."

None of my brothers or sisters went to high school or college either.

In my adult life, I have made a sacrifice so that all my children, both boys and girls, could go to high school and college. Every single one of my kids got their associate's degrees and several went ahead and got their bachelor's degrees as well. Out of all my girls, we now have a nurse, a secretary, some teachers and some mamas. My boys are now teachers, soldiers and policemen. I worked really hard to give them the opportunity to succeed, and it has paid off.

When I got tired of washing for the Silvas, I quit and I went to work in a tobacco factory for two and a half years in San Miguel. I worked there for almost three years but I had to quit because the smell of the tobacco was causing me nausea and headaches. I was puking and having headaches; I could not work anymore. All that toil and trouble was not worth the measly three Belize dollars a week that they paid us. Working at the tobacco factory, I had to wake up at three o' clock in the morning to cook breakfast and lunch for myself and my mother and all the smaller kids before I went to work. I did all that work so that when my mother and the other kids woke up they could just sit down and eat. I walked three miles to San Ignacio and got picked up in a little pickup truck and taken to the factory. After a few years, I had had enough.

I then went to apply for a job in San Ignacio hospital and I got it! There I worked as a "war maid." We mopped the floors, we checked pulses, we checked IV's and we made the biscuits for the patients' suppers. We worked in shifts. I did my war maid duty for three years. We were paid eleven dollars every two weeks. It was a little bit better. This was also a hard job. With my pay, I had to pay my rent in town. Rent was eight dollars a month for a small room and one or two meals a day. Thank God, at the hospital they fed us as well.

I met my husband while I was working at the hospital. We courted for three years and then got married. Back then, I thought getting married was the best option. I didn't know that being married meant a lot more work. When the kids started to come, I said, "Oh boy, now here comes the real hard work."

My husband and I lived on his father's farm on the opposite side of Chaa creek, a little farther up the river. I didn't like the way the in-laws were treating me, and he was not treating me so well either, so I decided to leave. After I had my first baby, I came back home to my parents and I never went back there again.

When my husband came to get me and take me back to his parent's place, I said "No! Only when I am dead will I go back to that place."

He said, "If you won't come back and live with me, then we will have to split."

I told him, "I love you, but I won't stay with you under those awful conditions in your parent's home."

Living with in-laws in a small house is no joke. There were fifteen people, including my husband and me, living in that small little house. It was all congested. I tried to make it, but for me, it didn't work.

I tell all my kids never to go live with their in-laws. I say to them, "You can get an apartment and get your own space. Don't come and live with me when you are married. Maybe you could live in the little house down the hill, in your own space, but not here in this house. Many things I do, your spouse may not like. It's better for you to live on your own."

If a guy wants to marry one of my daughters, I tell him, "Find a house and take your wife there. I am not going to bother or disturb you."

When people get together, I believe they need to be together by themselves. They can always visit their parents for a small portion of time.

When my husband saw that I wouldn't go with him, he was angry with me, so I stayed with my parents for three years. After three years, the owner of the village came to me and showed me a little house right here where I now have my downstairs kitchen. It was full of holes. He said I could live there if I wanted to. So I moved in.

When it rained I had to put a piece of tin up so I wouldn't get wet. There was only enough room in the house for the beds. My husband came back and lived in that little house with me. All the rest of my kids were born in that casita.

When Zena, my youngest, was ready to be born, my husband was building this big house which I am living in now. He wanted Zena to be born in the new, big house, but he was a bit silly and he hadn't built any stairs going into the house. He wanted me to climb a ladder to get into the house.

I said, "No way! I am not going to climb that ladder." I was pregnant and weighing almost three hundred pounds and he wanted me to climb a ladder? No way! So Zena was born in the little old house all full of holes in August.

He finally finished the stairs on the fifth of December 1988 and we moved into the new house. He had done a good job. The big house was very nice. He ripped out the old house which was made of pine. Termites really like pine. You could practically push the old house over with one hand. It took a long while to build this new house, and here I am still living.

My first born son is named Edilberto Leonel Waight. My next boy is Nary junior. My last boy is Abimeal. The girls, oldest to youngest are Thelma, Marlyn, Berta, Judy, Jeanette and Zena.

Even though I spent a lot of my life just trying to make ends meet and doing the work of a daughter, a wife and a mother, I knew that healing was my calling. I knew this from a very young age. When I was young, I would go with my father to collect plants and copal and learn little things from him. I learned from watching him. Because I was a girl, he didn't approve much of the fact that I was the only one who was interested in healing. He wanted one of the boys to learn, but they were not very interested. I wanted to go with him. I knew that healing was my calling and I wanted to learn more and more. Back then, people didn't recognize that girls could do good jobs. Of course we can do good jobs, just like the men! We can do it!

Later in his life when my father saw how interested I was in healing and how I was learning well, he said, "Beatrice, you are the only one of my children who will follow in your grandmother's footsteps. All the grandmothers on my side were midwives and all the grandmothers on your mother's side were midwives and now you should follow in their footsteps." He had given me his blessing to take up healing, but only as a midwife, not as a spiritual healer and primicia ceremony leader like he was.

Midwifery was fully available to me. From both sides of my lineage, there was something for me to follow. All my grandmothers knew what to do in many situations. A long time ago, my Grandmothers lived simply in the jungle and they didn't have anything like we have today, and yet they faired very well. They knew how to use massage, herbal remedies, women's ceremonies and prayers to help women through all the stages of life.

I learned many things from my mother and grandmother. It took me eight years to learn *la sobada* from my grandmother. *La sobada* is a very specialized massage for women's bodies. It is done on the abdomen and used to improve women's health, no matter what stage of life they are in. I wanted to learn la sobada, so I stuck with it with determination and little by little, I learned.

My first test was to go and find my own uterus by massaging my own belly. For three years I could not find my uterus. I knew I had one because I had been pregnant. I got mad at myself because I couldn't find it!

My grandmother said, "You have to find your own uterus before you can find anybody else's uterus."

During that time, I had a next baby. After the baby was born, I tried and tried and finally I found my uterus.

I said to my Grandmother, my teacher, "Okay, I passed my test. I have found my uterus."

She came closer and I told her exactly what position my uterus was in and she came and kissed me and said, "You are right." After that she trained me thoroughly. She gave me more techniques such as the pregnancy massage, how to turn babies from breech to head down and what to do if a miscarriage is just starting. My mother, who also knew these things, taught and trained me as well.

I also learned a lot when I was working as a nurse, a "war maid" in the hospital. It was my job to check the pulses of the patients and take their intake story. Through working in this way, listening to people's stories and taking their pulses, I learned how to identify the different spiritual ailments in the pulses, such as grief, *susto*, evil eye, envy and *pesar*. My father confirmed with me that I had taught myself the pulse reading correctly.

I birthed all of my kids at home, except for one. The oldest one, Leonel, I had in the hospital in San Ignacio. It is the law in Belize that your first baby must be born in the hospital. All my other babies were born at home. Home birth for me is better because you have privacy.

When I was bearing children, the midwives knew exactly how to help me. I had all those kids and I never tore once! I never had anything go wrong. Even having the breech baby, Berta, right here at home, nothing went wrong.

I think what happens nowadays is that women get cowardly. Birth is not about getting cowardly. Birth is a special time to be strong and bring forth that baby! If you get cowardly and act like a little chicken, thinking you can't do it, then, you will probably end up with a cesarean. I'm not saying cesareans are not good. Cesareans are very important and life saving, but only when you really need it.

La sobada, has a vital role to play in pregnancy. I only had one difficult and long birth. That was Berta and I know why it was so hard. Berta was breech because I had not received la sobada during my pregnancy with her. I didn't get any massage during Berta's pregnancy because I was hiding the fact that I was pregnant from my grandmother. I thought she might disapprove of me being pregnant again so soon. Only one year and three months had passed since my last birth. The two babies were too close together and I knew it.

My Grandmother would always say, "A woman should leave at least two years time in between her pregnancies to allow her body to recuperate and become strong and nourished again. If a woman gets pregnant too soon, then she is not taking good care of herself!"

I was worried that she would think I wasn't learning well or that I wasn't taking care of myself so I kept the pregnancy a secret. Because of my hiding, she didn't give me la sobada, her special massage on my uterus. Because of hiding and receiving no massage, Berta's was a difficult birth!

Thank God, with all my other pregnancies, I had nothing to hide and so my grandmother massaged my growing pregnant belly from the third month of pregnancy onwards and everything went perfectly. I drank my special birthing tea, walked around during my labors, got my hot belly compresses, stayed vertical and I experienced zero complications. My longest labor, besides the breech one, was two hours. Some lasted only one half hour!

My grandmother, who taught me the massage and the prayers, was a midwife for sixty years. When she was 100 years old, she stopped being a midwife saying, "I have done my mission. Somebody else has to do it now."

So, when my grandmother stopped working, my mother and I chipped in a little bit to help the birthing women, but when your calling is not to be a midwife you really don't like it. It is not my calling to be a midwife. I am a massage therapist (*una sobadera*), an herbalist (*una herbalista*) and I lead ceremonies, do spiritual healings and make prayers for people.

To my father, midwifery was the only acceptable healing path a woman could take, but I am my own person and have followed my own unique calling which combines the wisdom from both my male and female ancestors. I did midwifery a little bit, I still do it a little bit but my calling is not to be a midwife, mostly because I am a sleepy head. Babies like to come in the night and you can't say, "I want you to have your baby right now, so I can go to sleep!"

Babies come whenever they need to come and I like to sleep at night, so I just stick to doing massage, spiritual healing, herbal medicine, prayers and ceremonies.

About twenty-five years ago, I started my healing practice in earnest. This is how it happened. A lady in my village had four kids who had died and she only had one baby left. For several weeks, every week, she had buried a child. I don't know exactly what happened to them, but the last one, who was only one and a half years old, also fell ill. The grieving lady came to me with the baby early one morning. My father and my grandmother had just died. Everybody in the village had relied on my father and grandmother for healing but all of a sudden they were gone. First my Grandmother died and then three years later, my father died. At that time, I was only doing healings for my close neighbors and for my kids, but not all the needy people of the village.

One day, this sad woman who was named Miss Ek, showed up at my house saying, "I want you to heal my baby!"

I said, "How can I heal your baby? I don't know what to do!"

She said, "Oh yes you do! Your father was very sharp! He was a very powerful healer and I know that you learned something from him so please do something for my baby because he is dying."

The doctor had given up on the baby and had said to let him die. I looked at the baby and saw that it was grief that was killing this baby. He had a big head and a big huge stomach but his legs and arms were skinny. He had a black and blue patch on his bottom. That patch is a sign of the last stages of the spiritual sickness that we call grief or in Spanish, *"pesar."*

"What are you feeding this baby?" I asked her.

She said, "Coffee."

I said, "You can't feed a baby coffee! Give him some soup, banana and orange juice."

She was afraid of giving him anything because he was so sick. I told her to roast a chicken and give it to the baby to suck on. Then I said the prayers over the baby with my plants and copal. I did massage on his feet, gave him an herbal bath with prayers and sent a quart of special herbal tea home with her to give to the baby. The tea was made from garlic peel and marigold flowers.

I told her, "Boil it, strain it and anytime he gets thirsty, give it to him."

I told her to give him just a small quantity because he was so dehydrated from the diarrhea that he would only be able to tolerate it little by little. All the poor little guy had been stooling was something that looked like mud and water mixed together.

I told her, "Do all this with faith and if he makes it tonight, tomorrow morning you can bring him back to me at eight o' clock."

The next morning at eight o'clock, I could hardly believe my eyes! Here comes Miss Ek, laughing and smiling and carrying her little baby in her arms! She was so happy because her baby had made it through the night and was getting better! We continued the prayers, baths, special foods and tea for nine days and the baby lived.

When all the healings were done, I begged her, "Miss Ek, please don't tell this to anybody because I'm not doing healings right now. You see, I have all these kids plus the chores, the cooking and the washing. My time is very limited. The kids go to school. I have to feed them and get them ready every day. Then at noon, they are here and hungry for lunch. I have to haul my water from the river and work all day just to get by so, I really cannot do this! Please don't tell anyone Miss Ek."

Then, wouldn't you know it, three months afterwards, people started knocking on my door. I'm thinking to myself, "I am cooked already!"

I said to them, "How come you are coming to me?"

"Oh," they said, "Miss Ek told us!"

I felt sorry for them, I couldn't turn them away and I got myself into a bad situation because the washing was not done, the dishes were piling up and little by little I got myself engaged in this healing work. I did more and more until I got really into it! And still I had the house and kids to tend. Even to this day I am doing healing work.

Even when I need a rest many people come asking for help. I say, "No I am tired, I cannot do it!"

But, they are desperate and say, "Please just say the prayers for us."

So I say, "Okay," and burn some copal and say the prayers but I say "I cannot do the massage on you today, I need the massage myself right now!" After a while, I feel better and start to do massage again.

Soon after all those people started showing up at my door asking for healings, I had a dream vision of my father coming to tell me that I needed to take his place. In the dream, I questioned him, "How in the world am I going to do that? I can't do it with all these kids and a husband; I just cannot do it!"

He said, "If you cannot do it, I will have to take away your husband!"

I thought to myself, "How could I survive with nine kids and no husband?"

So in my dream, on bended knee, I promised him, "Please don't take away my husband and I will start helping people!"

In response to my dream, I started lighting candles on my altar to the nine benevolent Maya spirits. On the fourth day of my candle lighting, I had another dream vision. In my dream vision, a big board appeared and on the board was clearly written everything I needed to do. I also knew many of these things from watching my father. Now I knew that I was getting initiated with his approval and his blessing even though I was a woman. On the dream board was written how to make the herbal baths, how and when to burn the copal, what kind of copal to use for different circumstances and all of his special wisdom. There it was, everything I had wanted him to teach me when he was alive, written on that magical board.

My father was there, too, in the vision saying, "I'm not going to leave you. I am going to be always with you to help you whenever you really need help. It's not that you are going to do this by yourself, I am going to chip in to help you."

So, in the dream, I agreed, "Okay, then, I will do it."

That's how I started my healing work, with a promise that I made to my deceased father. Even though he was dead, I knew this was a real dream vision, a true message directly from him. All the things that I needed to do were manifested there on the dream board.

He also told me in the dream that I was going to get a *sastun* but that it was not yet time for me to get it. A *sastun* is a sacred stone or crystal used by Maya shamans and healers for divination purposes.

I proceeded to pray for a sastun for five years and nothing happened. Finally, one day when I was taking a nap in my hammock on the veranda under the house, I had a dream vision that showed me a certain spot, near my home, where lots of cassava was growing. The message I received in my dream was, "Dig right there!"

By this time I had just about given up on finding a sastun. I had decided, maybe it's not for me to find a sastun in my lifetime. I didn't understand the dream vision about digging up the cassava but I was going to do what I was asked. So I asked some kids to help me dig up the cassava. I thought maybe I had too much cassava and maybe I should share it with the neighbors.

I was thinking, "Is this what the dream vision is telling me? Do I need to share my cassava?" On the first day I went near the spot they were showing me in the dream and dug up some cassava.

We had two full buckets of cassava and the boys helping me called, "Auntie, are we going to dig some more?"

I called back, "Yes!"

I was doing my best to follow the message in my dream. We still hadn't dug in the exact spot that I had seen in the dream, but we were close and we brought up all that cassava. We sent some to my brother, some to my mother, some to my sister and some to my neighbor. I gave so much cassava away that day!

That night, I had the same dream and heard, "You did not hit the spot!" They showed me a little mound in the dream and said, "Dig deeper".

So, oh boy! I was a bit confused! I was thinking, "What is this message?"

I didn't know what the dream was trying to tell me. But I was doing my best to follow instructions. The next day, I took the fork and the bucket to the cassava patch again, and while I was walking down there I was thinking, "Maybe there is someone else who needs some cassava! I know what I'll do. I'll go to all my neighbors and call to them, "Okay, who needs some cassava?"

But I didn't have to do that because this time when I went to the exact spot that I had been shown in the dream, I dug once and got nothing. I dug again really deep, and as soon as I set the fork in the ground, something went POP! Up popped a round little stone!

"Oh!" I said out loud, "Now I get it!"

So that's how I got my sastun.

My father had said, "When you get your sastun, don't just grab it! Hide it somewhere and surround it with skunk root for nine days and if it is still there on the tenth day, then, and only then, is it yours. If that sastun is not yours, it will be taken away from you. Don't just take it; that's not good. Play like you don't want it and you will find out if it's really for you! If it's really for you, the benevolent spirits will not take it away."

Remembering my father's words, I put the stone in the downstairs kitchen and surrounded it with skunk root. Sure enough it was there on day ten, so it was truly mine. I did what my father taught me and on the tenth day, when I went to check on it, it was still there and I have it still.

It's a small little stone. If you hold it in your hand, it will give you guidance about a patient. For example, if you want to know if a certain lady will recover from her illness, it will tell you the answer.

When it gets hot, it is getting ready to give you an answer. If you see a coffin in your sastun, it means the person in question is going to die very soon. The sastun does not lie. That's why I don't use it. Knowing things like that can be really scary! The sastun tells you some things you should not know. People cannot always handle the truth. I don't want to be the one that gives those kinds of messages to people! My father didn't like knowing those things either. Some things are best not to know. It is good to leave some things as a mystery.

Dream visions are better. I work more with dream visions. Also, I want to help people, not scare them. I would never want someone to scare the hell out of me, so why would I want to scare the hell out of someone? That's why I don't use my sastun. I just hide it away. Before my father died, he gave all his sastuns back to the earth. He didn't give them away to anyone because everyone needs to find their own. You pray for one and you get it on your own. If it was just handed to you, you wouldn't appreciate it because there was no sacrifice involved. When you do your prayers, you get your dream vision and then you get your sastun, in that order. Whatever you get through hardship, you will appreciate more. We need to appreciate these sacred things.

Something similar happened to me with the healer's hut. Over a series of days, I was ripping out all my chicken coops and one night I had a dream vision about that area of land where I was working. In the vision, I heard this message, "Go right before six o' clock to that place."

The first day I didn't go, but the next day the same dream vision came back. I didn't go that time either. The following day, the dream came again, so I went there at six o' clock. That was the third day, and in the middle of the floor of the old chicken coop, I found a Maya face made out of clay- smiling and laughing from ear to ear! That was the sign that I had been told in my dream to go and find. Then I knew it was the spot for my little clinic. I kept that sacred smiling face hidden in a white bag under my bed. I could feel it always peeping up at me. But then, one day, it got lost! It disappeared after it did its job! Whoever gave it to me took it away once its job was done. It went back to where it belongs. I've searched and searched but it is nowhere to be found.

One time, a girl came here, who I think was a psychic, because when she stood in my healing hut, she said, "Miss Waight, did you find something sacred in here that revealed itself to you?"

And I said, "Yes!"

Then I told her the story. This is how I know that where my little healers hut stands is a sacred place!

I really wanted to build a healers hut right here on my property on that spot so I could have a special place to do my work outside my house. When you do healing work in the house, the bad energies that your clients release can fall on the people living there. My dream came true when fifteen women came to my house with Rosemary Gladstar.

They asked me, "Miss Waight, is there anything you want to make your work easier?"

I responded, "I want a little clinic because doing healings in the house is not so good. There is no privacy, and when negative entities are taken out, they can hit anyone in the house."

Those generous ladies all chipped in so that my little healer's hut could be built, and that is the one still standing there! I like my healers hut because it is not too close or too far from the house. I can have privacy there when I do my healings. When I say prayers for people, the music in the house distracts me. I like it quiet when I work. I get distracted if the radio is on. In this practice, you need to be focused. When I focus, I can really do some good work. If I am not focused, I get carried away and I get lost and confused, but when I concentrate, I can do a very good job. That is why my intuition told me that I need to have my own clinic, my own little healer's hut.

On my fifty-sixth birthday, my friends from Boulder, Colorado, brought me a massage table. Before that, some doctors and nurses that I was teaching saw that I was doing the massages on a bed and I had to bend over to do my work. After two or three massages, my back would be killing me. They made me a rough table that could hold one thousand pounds, I still have it. The other table that my Boulder friends gave me, the one I use now, can hold 300 pounds. So now I have a table that can hold a fat, fat person and a table that can hold a lighter person. That helps me a lot.

The things I have in my healing hut were all donated. The little table, the chair and the shelves were all donations. I think Spirit gathered these people and told them what to donate. Some people even came and gave me beautiful pictures. Way back, I had nothing. When I have all my tools together, I can do a better job. You wouldn't send a child to school without a paper or a pencil. You can't send someone to cut wood in the jungle without a machete. It's too hard to do your work without tools.

I have a little hammock in my healer's hut for my students to rest in when I teach classes. When no one is in there, I go there and find tranquility and peace, and I tell the kids not to bother me unless my mother or sister is dying. Anybody else can wait because I am going to rest. It's nice to rest all by yourself with nobody around to mess you up.

One time, after being in my healer's hut resting a little while, I felt somebody patting my forehead and touching me. I thought it was my youngest daughter Zena. I didn't open my eyes but I scolded, "Zena stop it! I am not playing with you." After five or ten minutes, again it was being done to me so, I said, "Zena, don't get on my nerves because you know what's gonna happen if I get mad!" It happened again so I opened my eyes. I thought she was under the hammock. I said, "Zena, this is the last time and I am warning you! If you come and do it again you are not gonna like what I'm gonna do!"

Then it happened again! Someone was patting me on my legs and my forehead and my arms and I went to grab the hand but nobody was there! Now I know the spirits just wanted to play with me. There are good spirits which make you feel peaceful and joyful and then there are the bad ones. When the bad ones are around, you can feel a cool chill come over your body.

One time, some people from Denver invited me to go to Colorado to teach a little workshop. I was a little bit shy and intimidated because I had never taught a workshop all alone. But then I started thinking, "Whatever you want to do in life, you can do it!" That's when I started going to America to do workshops, ceremonies and healings and teach about pulse reading, family wellness and women's wellness. I have been invited by many massage therapists and other practitioners to go to America and talk about what I do here in Belize. In my travels, I meet a lot of people that want to come and study in Belize. I have been many places: New Mexico, Arizona, Colorado, Vermont, California, England, Guatemala, Mexico and Honduras. I have been a guest speaker at The Women's Herbal Conference with Rosemary Gladstar in New Hampshire. I was also invited to do a speech at the Women's Herbal Conference in California.

I love to work with women. It is not that I don't like men. The men can participate, too. When I do family wellness classes, I think the men should come because the remedies apply to everybody in the family. I teach basic things about what to do if you have an ear ache, a tummy ache, a headache or a fever. For really serious concerns, I say, "You have to go to the doctor."

As a healer, I know my limits. I know what I can do and what I cannot do. I never say I know everything or can do everything. In this life, we are in the process of learning. Every day I learn little things. Even if I live to be one hundred years old, I will still be learning.

Once I was teaching a class to a group of doctors. One of the doctors from Greece wanted my email and phone number, and he put my picture and my information up in Greece, so that the people there would know about me. I didn't believe him. I thought to myself, "Ah, that's only sweet talk."

After a few months, I was really surprised when a lady from Greece contacted me. She had seen my picture and was asking for help. She was limping and her body was leaning over to one side. I asked her for her case history. She said she had a hysterectomy twenty-five years ago. I got the idea that it was scar tissue that was accumulating on one side and causing her to limp on that side.

She asked, "How many days will the treatment take?"

I told her seven days. She came and stayed with me in Belize for seven whole days. In five days, she could stand up straight and walk normally. I gave her prayers and massage, steam and herbal baths. She was so pleased about it!

One time we had a group of ladies from England come to Belize. One of the women's uterus had been hanging out (uterine prolapse) for twenty years.

She said, "No way am I going to have a hysterectomy!"

She had seen a brochure about me in New Hampshire and she came to Belize to get some help. She asked how many days it would take. I told her seven days. She came and I worked on her. By way of the Maya treatments, her uterus was lifted up and stayed in. When she got back to England, her gynecologist was so amazed that he wanted to meet me. I thought, "Oh no, now I am in trouble!" But I prayed about it. I said in my prayers, "If it is a good thing to meet this man, let it be so, and if it is not a good thing, then please don't let it happen."

Two days before I went to England, the doctor went to India so I didn't meet him. It was not meant to be. We only saw each other in pictures. I went to England to teach workshops. They sure eat a lot of potatoes over there!

I am the last one to fully practice our Maya spiritual ways in my village. My family members split from this way of life even though they were brought up in our culture. For what reason, I really cannot tell you; they just forgot about the sacred ways that our father taught us.

They don't even burn copal anymore. They believe that copal is evil, because the Evangelists who came to our village brainwashed them. Some of my family members knew all about our Maya ways, but the only thing they do now is to plant herbs and collect herbs, without ceremony, burning copal or even collecting copal. They will not practice our ways anymore because of the Evangelist Church. Some of them are afraid they will be teased for doing witchcraft, but I don't care what other people say about me; I do it anyway because it has heart and meaning for me. You can never suit everybody. I would die trying, so I suit myself! I feel good about what I do. People can be mad until they are blue, and that is their problem, not mine. I know who I am and I am proud of whom I am. I don't hide who I am. My conscience is free and clean and to me, that is what is important.

So what will happen to all the kids? The kids could just go astray and this rich, beautiful culture could be lost for good. I am the only one.

One of my family members went very far astray. He foolishly went and had a child with another woman while he was married, and he knows that's not proper, so he can't do ceremony anymore. You see, some exclude themselves because of what they have done. They chose that path. It is sad. Then, when they see the benefit of what this path is giving to me, since I still practice it, they envy me. What you choose in life is what you get. The old group has stopped doing Maya ceremonies. When my father was living, he was a guardian angel, keeping everything right. When he died, everything fell apart. He was holding it together. You see, my mother, up until the end of her life, still believed. She did her prayers and burned her copal, but still, something was missing ever since my father died.

When my ancestors lived in the Yucatan, there was a Catholic church there. The Saint they honored there was Saint Joseph. In San Jose Yalbec, there also was a church in honor of San Jose, but it was burned along with their houses. My ancestors practiced both Catholicism and their Maya spirituality. For them there was no conflict between the two traditions. In fact, they complimented one another quite nicely. When my parents and their little group first came here to Santa Familia, there was no church here. There were only seven families, but little by little, they built the village and the church. The church building was also the school and it was dedicated to the sacred family. Even with the church thriving, we still practiced our Maya ceremonies alongside our Catholic faith.

There are many reasons why Maya spirituality has faded away in our village. The Christian evangelists came about twenty years ago, and with them came trouble. They would offer us clothes and money and they would try to hook us into their deal. When you have a membership to that church, you have to go to church three times a week and they do a collection for the Pastor. The Pastor doesn't work outside the church, so the collection pays for all his food, housing and bills. The evangelist church members are still working for the things they received as "gifts" so long ago. Now it is payback time. It's a never ending pay back. The evangelists draw you away from any family members who are not converted to the evangelist beliefs. They separate you from your culture and your spirituality and tell you not to associate with anyone outside their church because people outside their church are considered "not holy enough." The evangelists spread the lie that burning copal and practicing Maya ceremonies is the work of the devil.

Electricity also came to the village about fifteen years ago and that changed a lot of things for us because with it came television and radio. We also now have different foods available to us that are modern but not so healthy, such as ramen noodles. Many people in the village have abandoned our traditional diet for the new foods because they are easier to prepare. Thank goodness, many of my family members and people in the village still grow beans and corn and we all celebrate special occasions by making our traditional foods like *tamales* and *bols*. Some of the villagers have not been converted to evangelistic Christianity and maintain their Catholic faith, but they have forgotten about our Maya spirituality and healing, which is so powerful and sacred. It is a huge loss.

I would love to pass on what I know to someone in my culture, but anyone in my village who I might teach is too young, as of yet. An apprentice has to be mature to take on the responsibility of practicing Maya healing and leading Maya ceremonies. It is something powerful and sacred to hold and a person has to be strong and wise to do it. It is a serious responsibility, not something to be messed with. Because nobody here is ready and few are even interested, I have taught some special women from outside my culture with the hope that they will carry on the Maya traditions with respect and care.

I keep on practicing my Maya way of life and try to pass it on a little bit to those who are interested. I do my best to keep the promise I made to my father. Keeping my promise and carrying on my tradition has been a blessing to me. Because of my healing work, I have traveled a lot and met tons of people. I have friends and students all over the world and many of them come to visit and learn from me still. I love what I do. It's not always easy, but it's a lot easier now without little kids and a husband to take care of. My kids are grown now and they help me out a lot.

I was born right here in Santa Familia. This is what I call my paradise. It's not that there aren't places more beautiful, but I am so used to my place. So that's why I call it my paradise. Every morning I can hear birds singing. I see wild hens and I am surrounded by one hundred fifty healing plants, some of which I planted and some of which are just growing wild in the jungle. This is what I love and I think I'm going to die right here. Until then, I'm going to continue to do my practice.

Now I am going to address a Maya prayer which I have translated into English.

"Great holy spirit, wind spirit, as we stand here this evening, we ask you for healings and blessings for our family and friends, for good health, for guidance and for keeping us safe every day. I ask you in the name of the great healer, *Ix chel,* to listen to my prayers so we may live in peace. Amen."

Maya Plant Ally Number Three: Datura

Datura Aborea

Datura (trumpet flower) in Miss Beatrice's yard (photo by Katherine Silva)

Datura is a very powerful plant that is not for messing around with. It is never to be used internally because it is toxic and hallucinogenic. However, you can soak datura leaves in rubbing alcohol and then use it as an external liniment for all of your aches and pains. Datura is really good for a persistent cough. To rid yourself of a stubborn cough, rub castor oil on your chest and then place a big datura leaf on top. The leaves will cook on you and become crispy. That is how you know the cough has been pulled out. Datura is also good for asthma. Put coffee grounds, in the shape of a cross on a datura leaf and then stick it to your chest with olive oil. Do this every day and the asthma will disappear. For eye care, for the advanced herbalist *only,* boil one flower in one quart of water and strain out and save the water. Put two drops of the water in each eye to make your eyes "happy and healthy." You can put olive oil on a datura leaf and set it on a sprain. A poultice of mashed leaves is good to apply to an inflamed tailbone or sacrum. It is very important to pay attention to the doctrine of signatures and heed what it is telling you. See the way the datura flowers are hanging down? This tells you that you cannot trust datura because if you drink it, you will go down like those flowers, so don't drink it! Don't even put this plant in your bath! None the less, datura is an important herbal ally and I have a big beautiful datura plant in my front yard. We call it "*campana blanca*" (white bell) or trumpet flower, and we treat it with respect. Other names for this powerful plant are jimsom weed, *xanica* and *floricundia.* In the United States you have a small datura plant just as dangerous and just as powerful. You can use those leaves externally, the same way we do here.

Chapter Four: Maya Healing for the Aching Soul

Neroli Blossom, Yucatan, Mexico (Photo By Katherine Silva)

"Once I have read their pulse and listened to them properly, I burn copal and I say prayers for them into their pulses, their forehead and heart with my plant allies; basil, marigold, rue, rosemary or white roses."

I am now going to share with you some of the teachings that were passed down to me through many generations in my family. My people, the Yucateca Maya, have been practicing these techniques, these prayers and these ceremonies for hundreds of years. The women in my family, all the way back, practiced these powerful Maya ways and they were all midwives. The same goes for the men; they were all healers. So that is how this was passed down to me.

The Maya believe that many physical problems are caused by spiritual ailments. There are many different types of spiritual ailments that we find in people by reading their pulse. The pulse is an important way to diagnose grief, shock and sadness. My grandmother never went to school, but she was an excellent pulse reader. My father only went to school for two years and he was really good at pulse reading as well. I learned a little bit from observing my father, but I learned even more when I was working in San Ignacio hospital for three years. I was the one assigned to take pulses, temperatures and case histories from the patients. After three years of practice, I had a much better idea of how the different pulses feel.

When I finished working, I came and told my father that now I could detect a grief pulse and he said, "Yes that is true. You have it right!" I learned pulse reading on my own and that served me well. Learning pulse reading takes a long time. You have to do a lot of hands on study and listen to many case histories before you can practice pulse reading with confidence.

Pulse reading is a very important part of the Maya way of healing. When someone comes to me for help, I read their pulse and ask them what has happened in their life, all the way back, even during childbirth. Maybe they were born with forceps or a c-section, and there is a little bit or perhaps a lot of birth trauma that goes on and on in their life. These things are important to know. I do not force them. If my client wants to tell me, then that's okay; if not, that's okay, too.

When somebody comes in for a treatment, I am very polite and treat them as best as I can. I offer some water or tea or offer the bathroom to break the ice. If they come directly from the sunshine or a busy day, the pulse I get at first may not be accurate, so I give the client a little time to rest and then I start. Sometimes the person will tell me every little thing that is happening. Sometimes the clients don't say anything. If I think they don't want to be questioned, then I don't question them; if they don't want to say it, they are not ready. I just go ahead with the treatment.

Women who have been sexually abused, raped or have experienced incest may not want me to touch their abdomen or solar plexus, because it brings up a bad memory. In that case, I massage the feet. I give a really good massage really deep in the sole of the foot. Then we sit and talk, and with much emotion, they may tell me what happened. I give them some blue vervain or chamomile tea. What is really good for these women is the neroli water. If you have an orange tree nearby, that is really good, because that makes them feel a little better. I pray into their pulse with neroli, give it to them to sniff or put it in their plant bath water. We call sweet orange blossom *asajar*. My grandmother often collected the blossoms and made them into perfume by soaking them in oil or water. She also used her homemade neroli blend for sprinkling around the house.

Once I have read my client's pulse and listened to her properly, I burn copal and I say prayers into her pulses, her forehead and her heart with my plant allies like basil, marigold, rue, rosemary or white roses. I do massage on her solar plexus and her feet.

If my client is a baby, I massage the feet with mild oil like virgin olive oil mixed with a little bit of rose or lavender essential oil. It has to be a type of oil that causes no problem to the baby. I mix essential oils with olive oil. I do not use straight essential oil on a baby. That would be too strong. Olive oil will not hurt anybody, not even newborn babies.

Depending on what my client needs, I may make a plant spirit bath for her, recommend special teas or herbal formulas, do an egg clearing or recommend a "letting go" ritual. Even after the first treatment, the person or baby will feel much calmer.

Being a healer, I have to work with plant spirits, a deity and the nine benevolent Maya spirits. I have to call in my plant allies and honor my deities by lighting a white candle and making them a little altar where I can hold a little statue of a saint I am dedicated to. I like all the forms of Mary. I like Guadalupe and The Sacred Heart of Jesus. Those are the ones I like so much. So I offer to my deities a white candle, some white flowers and also I say the *Lord's Prayer* and the *Hail Mary.* After I say my three *Our Fathers* and my three or nine *Hail Marys,* I say my Maya prayers. For me, it works.

You also can make a little altar, a little space, where you can put a statue of a deity that you like. You can offer them a white candle and white flowers and say a prayer that has meaning for you. Any prayer that you know, and is meaningful to you, will work for you.

My prayers may not work for you. You can pick the ones that you normally say and that have meaning for you, but if you want to say my prayers, go ahead, I can share. I use the *Our Father* and *Hail Mary* and sometimes I use another prayer that my father taught me. Whatever prayers you pick, you must have faith in them. If you do not have faith in your prayer, it will not work; it will be useless. Faith is what makes things work.

As a healer, I need to work with plants. I don't just go out picking plants recklessly. I say a prayer to each plant, when I pick some of their leaves, to thank the spirit of that plant. You could do this also, not in my way, but just in your own way. You need to offer a thanksgiving prayer. As long as you give thanks to the plant allies, they will come and do the work. When you do your plant collecting prayers and you believe that the prayers will help, and that the plant allies will come, then a true healing can take place.

Here is how I pray when I collect my plants,

"I am the one, Beatrice, walking thru the hills and valleys by the creek, by the river collecting plants for______ (say their name or just "people" if you don't know who will be using the plants) to heal their __________. With all my heart I believe it is going to be so. I give thanks to the spirit of this plant in the name of the Father, the Son and the Holy Spirit, Amen."

Just whisper the prayer to each plant as you collect them. If you buy herbs in the store, hold them with your special intention and your special prayer, because you don't know if a prayer was said when they were picked. There is no problem saying a next little prayer. You can put your own good intentions into the plant. There is no such thing as saying too many prayers.

As you pass by, you can see your plants waving at you so you know there is a connection. People who do not believe these things will just laugh at you. They will think you are *una loca,* (a coo coo). But it happens. It really happens! The plants know who you are. They recognize you and so they say hello and call to you to pick them and use them for medicine. Plants love to come and help. They love to be treated with respect, not just to be picked without recognizing that they are here to be helpers! Plants love to be called and used.

The plants are here to be helpers. My grandmother always said that when something happens to you, walk 300 feet and there is the plant for your medicine. What happens sometimes is that we don't notice the healing plant. We are just kind of blind. We might mash it, cut it, or burn it, and maybe that plant was our medicine right there where we were just tampering around. If you say your prayers, you will always have dream visions and the plant spirits will come and talk with you and tell you what you need, and then everything is going to be okay.

Copal is a true part of every treatment. I burn only black copal, not white, to ward off bad energy. Black copal can be mixed with garlic peel and dried rosemary. We never mix black copal with white copal because white invites and black dispels, and mixing the two together is as if to say, "Come here and now go away!" That would be disrespectful to the spirits. White is for good luck, new beginnings, new projects and blessings and should never be used to ward off anything. Clean your copal burner really well to get the black copal ashes out before you burn your white copal.

Talking more about copal is something I cannot leave out. My father told me that copal is so sacred that it is called "the blood of the benevolent Maya Spirits." Every kind of copal needs to be treated with very special care. The different kinds of copal come from different kinds of copal trees. Every kind of copal is sacred. Burning black copal wards off evil. If we need a loan or a blessing or something good to come our way, we burn the white or the grainy or amber kind of copal. The light colored kinds of copal attract blessings to us.

We Mayas have another way of protecting ourselves in our daily life. We wear amulets. Amulets are something we wear on our bodies for protection. An amulet is sewn on the full moon. So, if you sew it on the full moon, it is good for three months. Then we will make a new one and discard the old one. The old one is burnt when it is finished. The other thing is that the amulet is sewn in black cloth with black thread and everything you put in the amulet is sacred. You take your most precious stone, wash it and leave it over night in the full moon and rinse it. The full moon time period is three days before the full moon, the full moon itself and three days after the full moon. Other little things you can stuff in the amulet are your herbal allies. We wear the amulet close to our hearts or in our pocket. Everyone in the family needs to sew one for their own selves. The amulet dispels negative energy away from you and protects you.

Any type of spiritual malady will get better when I do the prayers in the pulses, make an herbal bath and burn copal. But for each spiritual ailment, there are also specific things that I do.

The first spiritual malady that we are going to discuss is commonly called "evil eye," which in Belize we call *mal de ojo*. *Mal de ojo* usually occurs in a baby who receives too much attention from strangers.

It happens when lots of outsiders are admiring the baby, saying, "Oh how cute!" and showering the baby with constant attention and praise.

It's too much energy for the baby to receive and this causes what we call "evil eye." To avoid this problem, we cover our babies up when we take them out.

When babies have evil eye, their stool might turn green. They may have diarrhea and be very irritable. Here in Central America, people talk a lot about evil eye. Evil eye is not intended. It's simply too much admiration sent to one person. When it's a baby, a mother's love will not give evil eye. It usually comes from a stranger or someone the baby doesn't know very well who is admiring him without touching him. It sends a hot energy to the baby, which is too much for the child. Evil eye could even kill a baby if it is not treated properly! That is why we protect our babies and treat them promptly for evil eye. Evil eye can also occur in older children and even adults. The symptoms are usually similar but not as severe and still need treatment.

The first symptom of evil eye in a baby can be vomiting or green diarrhea. Often, the body is cold and the head is hot. The heat concentrates where the admiring energy was focused. That is why the face is often red and overheated. Babies have cute faces so their head gets hot with all that attention. That is evil eye. We believe babies are very vulnerable to evil eye until they are one year old. After they reach one year of age, they can go where there are lots of people, but before that it is better to keep them home or keep them covered.

Let's talk now about the treatment for evil eye. If you are the one who gave evil eye to a baby you don't know by adoring him too much without touching him, the best thing you can do to cancel the negative energy is to pat the baby on the head. Patting the baby's head takes evil eye away if you were the one who gave it. To protect a baby or child from evil eye, we use red coral on the left hand or dress the baby in red when he goes out and is exposed to lots of people. The red color cancels everything. When a baby with evil eye comes to me, I place basil, marigold and rue in cross form on the pulses, the head and the heart and whisper my prayers. With those same herbal allies, I do an herbal bath for the baby. Then it is time to burn black copal, just a little pinch. If the mama is sitting in a chair holding the baby, I burn the copal under the chair. The mama and the baby get a little bit of copal. They will need to come for this healing at least three times. The mama can bring the baby seven or nine times if the mama wants to, but three times is more than enough.

The next spiritual ailment I want to talk about is "fright" which we call *susto. Susto* is when you get a shock. For example, you are in a fire, car accident or some kind of trauma. The *susto* pulse is bounding and bouncing on the surface. Who can say they never had a fright? We all can get fright easily. Toddlers get fright if they are spanked in potty training or put to sleep in a dark room all by themselves. That kind of thing is not good because you can really scare children.

When mamas are pregnant, if they get a fright, the baby gets a fright too. If you get attacked in your house by banditos, if you see or are in an accident or if you fall from a horse, for sure you will get susto. Also, seeing a scary movie, watching someone be attacked by an alligator or experiencing a house fire will give you susto. When you have susto, you cannot sleep. You are restless and you have bad dreams which are often about your safety. For example, someone might be shooting you, drowning you, chasing you or pushing you off a cliff. I have even seen people who had anemia from untreated fright. The lack of sleep and loss of appetite lead to exhaustion and malnourishment. A body with susto is not happy or healthy. Many people with these symptoms check with their doctor, but the doctors do not usually find out what is wrong because no machine can detect spiritual ailments.

I treat fright with herbal baths, copal and prayers. I whisper the special prayers into the person's pulses while I touch them with the plants. I give solar plexus massage and some nerve tonics. A good nerve tonic for an adult might be tincture of man vine, valerian, St. Johnswort, Kava Kava or skullcap, but for a baby, passionflower or chamomile tea works well. I give plants that are good for the nervous system. If the person is a baby, I'm very careful. I do not give them anything too strong. Small quantities of a gentle tea are the best for babies. When they are five years old, they can start taking some of the gentle nervine tinctures in small amounts and that will help them a lot.

Now we go to a serious one, envy. In Belize, we call envy, *envidia.* Envy is when you live close to someone who sees you often and that person puts negative vibes on you. This is one to watch out for. I experience the effects of envy when I am doing my healing work and a neighbor or relative doesn't want me to do it. I might get a headache, unexpectedly break a glass, or feel a fever coming on. Well, that is envy. Perhaps you have a really close friend who knows everything about you, and she knows you went to Belize, so she's thinking "Oh she has tons of money and how can she take her whole family to Belize?" or she thinks, "How did she get that new car and the big house?" She doesn't know that you worked so hard for these things, and she is jealous! If the envy comes from someone in or near my house, when I go in, I don't want to stay there. Everything that I see in my house seems wrong. I start thinking, "Oh that picture doesn't look right there and, why is that cat there and why is that chair there"? I find any little thing to make me mad. I want to go sit in the park because when I sit in the park, I feel better.

It could be a close neighbor thinking, "How did she buy this big house?" If it's a new car that they envy, every day something might happen to that car. The first day, maybe there is a flat tire. The second day, it could make some funny sounds. On the third day, maybe you almost hit somebody. On the fourth day, it doesn't start. I'm not saying cars don't break down sometimes, but not every day! So, then you need to know that something really serious is happening.

It is envy, and most of the time envy comes from someone who really knows you. This is why it's not good to say from A-Z about yourself to everyone. You can trust people sometimes, but not always. Friends could be traitors! Maybe you are very responsible in your job, and some of the other workers were there before you, but they are not as responsible, clean and tidy as you are. Through your hard work and dedication, you prove to the boss that you are all these things, and much more, so he lifts you up to a higher post. When the co-workers hear about it, they are envious and send you bad energy and that can make you sick.

Symptoms of envy start with temporal headaches out of nowhere, feelings of weakness, exhaustion and sleepiness, achy joints and a fever that has no other explanation. One day, you might notice that you missed one hour from work and the next week you miss half a day. The next week, you miss a whole day and just gradually it gets worse until you are missing possibly a whole week. That is the intention of the people giving you envy. They don't want you there being better than they are!

If you think you envy somebody, you could do prayers. Instead of sending envy, you could give them good energy. For example, when I went to my friend's house in Arizona and I saw the beautiful rose bushes in her yard, I was aware that my feelings for them could actually turn to envy. I decided to say a prayer for her and send good energy instead. Rather than wishing I could actually have the rose bushes, I said a small prayer for my friend to have even more rosebushes. I think people who have envy and jealousy and don't stop themselves with prayers are evil doers. Some of my so called friends are just like that. This thing called envy is everywhere, not only in the US. It's here in Belize, even worse!

Another thing that happens with envy is that all of a sudden, almost every day, you break something. Glasses and plates will just happen to fall from your hand. Every day you break something, so this is how you know the envy is there, because normally that doesn't happen to you. Also you can't find the important things you need. Things hide from you. Thank God we have some ways to protect ourselves from envy and to rid ourselves from it, which I will discuss now.

What I do for envy is to burn a little black copal. I burn only black copal, not white, to ward off bad energy. Black copal can be mixed with garlic peel and dried rosemary.

After I treat my client for envy, I know that she will no longer be receiving negative energy from somebody who wants what she has, but who is too lazy to work for it. In this life, if you do nothing, you get nothing. For me, this life is about struggling and hardworking. If I want to get something, I can't expect that it just comes to me. I have to work for it, and I always treasure it more.

Another thing I do for envy is to buy an aloe vera plant and put it in front of my house. If that aloe dies, it means it took all the bad energy away from me, and then I take it far away from my house with thank you prayers and bury it. I also take a bunch of marigold flowers, really yellow ones, and put them in a jar in front of my house, and every other day, I replace it with fresh ones. A round mirror will work too. I Place it at the entrance of the door. When anybody comes in, putting negative energy on me, they will look in the mirror, and it will backfire on them! It is not a problem to protect myself in this way because I am not doing anything wrong. Saying my prayers, taking plant spirit baths, taking sea salt baths, burning black copal and placing marigolds and aloe vera by my door helps me so much.

What I also do is take a sprig of basil and submerge it in water on a Friday. The next day, Saturday, I sprinkle the water on the front steps and all the doorways of my house.

For my friends who work in an office where they can't burn copal, I suggest that they put an aloe vera plant in the room. Or, they can take three sprigs of basil, three sprigs of marigold and three sprigs of rue and mish mash the plants in the water with prayers. Then strain the herbal water and put it in a spritzer bottle. Then I tell them to go from corner to corner spritzing the herbal water in a cross form while whispering prayers nine times every day. That clears the space. Rose or lavender essential oil can be added to the spritzer and nobody will say that it is witchcraft. They will just think it smells nice and you are freshening things up. People who are not raised knowing about spiritual clearings and plant medicine, might not understand. When we clear our space of envy with a little spritzer, it just looks like we are making a nice odor. At home, I hang thirteen cloves of garlic braided together by the door, or I plant marigolds or life everlasting all around my house in a circle. All of these remedies help a lot and then we say goodbye to envy.

Grief is another spiritual ailment. We call it *pesar.* Grief comes from a great sadness or a loss in life. When we have grief, we feel heaviness in our chest, especially in our hearts. This is not a real physical heart problem but a big heart ache and a heaviness. If left untreated, grief does affect our physical heart as well. Grief is not going to kill quickly. It takes a long time, but little by little, *(poco a poco)*, it takes its toll. We get pesar when we lose something or someone that we were really attached to. For example, a divorce, a separation, or death of a loved one can cause pesar. Children can get grief if they are separated from their parents. A baby who was breastfed and then was weaned really quickly might have grief. Grief can also affect children whose mothers get pregnant again really soon after they were born and then neglect them, to a certain extent. You could get grief from losing a house or a co-worker. If you have to sell your house because things get so bad or a fire came and burned your house down and it was full of good memories, this could cause you grief. Maybe you couldn't save any of your photos or memoirs of all the good times or pictures of your grandmother or father or someone you are really attached to. Say, your little pet was smashed in the street and you were very attached to him. Of course you could get a next pet, but it's not the one you lost, so you have grief.

The first symptom of grief is sighing. Then it's followed by frequent coughing fits and a strange kind of wheezing. Someone with grief could get anemic because they lose the desire for good food. A grief-stricken person's hair may fall out. There could even be a lack of vitamin B or iron in the body of someone with grief. Skin conditions and numerous other things can happen to people and babies when they are neglected or they lose someone they love. Babies will start with crying and crying all day, and then they don't want their breast milk anymore because they lose their taste buds. A child with grief doesn't want to play even with his favorite toy, and that's a really bad sign. Grief can go on for years and years and you might not even know that you have grief. With babies, it can go on for about six months, and if it goes untreated, the baby will get a big tummy, a black and blue patch on his buttocks and the doctor will not be able to find anything wrong with this baby because the machine cannot see it! It's a spiritual cause for these physical ailments. If you are lucky enough to find a healer that knows about grief, the baby can get better again.

In the case of grief, when I do the plant spirit bath, I put a little neroli (orange blossom) water into the bath water because I want to create a beautiful aroma. I add red roses as well. If it is a baby and the baby is sleeping in a hammock, I tie a little red cloth where he can see it so he keeps looking and looking at it. Red is colorful and it gives hope and protection. On the baby's little shirt or dress, I put a little neroli water so she can smell it throughout the day and little by little it will help her forget what has happened.

Untreated grief can lead to cough, lung problems and allergies. If it's not real asthma, and simply a lung problem caused by grief, I do an infusion of garlic and honey. In one quart of wild honey, I put one to three peeled and chopped cloves of garlic and infuse it for nine days. I leave space on the top of the jar so I can shake the honey daily while it is infusing. I give this garlic honey to my client. If it's a baby or a small child, I give one teaspoon two to three times a day. If it's an older child or an adult, I give one tablespoon three to four times a day. Before it is gone, I prepare the next bottle. After two bottles of garlic honey, the lungs will have improved. I know in the US, you don't give honey to babies, so you could find another cure for them. Here in Belize we have a special wild honey that is totally safe for babies.

Also, in the case of grief, I make the plant spirit bath with annatto leaves because they are red. I gather nine leaves of annatto with prayers. See that big tree over there? We call it the lipstick tree. That is annatto. I take nine leaves of that tree, gently, while I say my prayers, and mix them with basil, marigold, and a pinch of rue or rosemary or St. John's wort. I make a nice bath with those plants. I crush all my plants into a bucket of water and sun it. The bath will be warm and ready in one to two hours.

If it's warm outside, I bathe the baby, or the adult, outside because the wind will carry everything that's causing illness away. Before the bath, I prepare black copal on a charcoal and make sure the person having the bath is in their bathing suit. Once I'm ready, I pour the bucket of water, with plants crushed in it, over them, one cup at a time, from head to toe, until all that healing water is gone. All the while, the water is being poured over the person's body, the copal is burning and I am repeating the prayers in a whisper. That person is going to feel much better.

If the spiritual ailment is really serious, my client may need a "let go" ritual. He needs to take nine pebbles and go by a river that's running, stand with his back to the water, and with his right hand, drop or throw one pebble at a time into the water saying "I want to let go of this grief, this fright, this sadness."

I recommend he does this ritual for nine days, or as many days as needed, until he feels free. When I do this ritual, I feel that something has stayed in the water and it's not on my shoulders bothering me anymore!

Sadness is like a small grief. We call it *tristeza.* It is a cousin to grief. It's why kids suck their fingers. It's insecurity. Maybe they were not breastfed, or perhaps they were put up for adoption. But, maybe by chance, the baby was treated with massage and prayers with basil, marigold and rue or rosemary for nine days. Then that baby is going to be happy again. If that was not done, he is going to be a troubled child, because for the rest of his life he is going to carry that sadness. That is why he sucks his fingers and feels insecure. When you hear that kids are troubled and then they start using drugs and thinking about suicide, sadness could be the cause. For them, life is not worth living.

For sadness, I make the same plant spirit bath with annato leaves, basil, marigold, rue or rosemary, for nine days. Also, I burn black copal in the room of the person with sadness. If the smoke makes him cough, I burn it in the room first and then put the person or baby in after it has aired out a bit. Someone with sadness will need also to do a let go ritual and they need a lot of love. If it is a child with sadness, older than two years, I give a belly massage as well but I'm very gentle.

There are other things that I discover when I do pulse readings, such as entities, spirit attachments and curses but these things are very serious to work with so I won't go into much detail about the treatment.

I will say this much however, if my client does have an entity or spirit attachment, I give them skunk root, but I do not leave the person unattended. When someone is getting over one of the more serious spiritual illnesses, they sometimes get very sleepy and tired or aggressive and agitated, so I attend to them and do not leave them alone. This is a separate form of healing. It is more serious and you have to know what you are doing. That's all I can say about it. Entities are best left to someone with experience in handling such things; they are in a whole different category.

Thank God the Maya way of healing has been passed down through the generations and we have access to it today. Way back in the rainforest, the people knew how to prevent and heal spiritual ailments.

I am now going to tell you how to make *un bajo sagrado*, (a plant spirit bath). First, collect all the herbs with prayers. Get a bucket that will hold two gallons of water. When you select your herbs, avoid the ones that have brown spots or white spots or big holes. Always look for herbs that are healthy. Now put all your herbs into the water with your prayers, and mish mash them or take a scissors and cut them up into small pieces. Put the mixture of plants and water in the sun to sit for one or two hours. If you need to, please take your soapy shower first, because we don't mix soap with our plant spirit bath. When the water is tepid or warm, this is how you bathe with it. You pour the water over you, little by little, with a small container such as a cup, starting with your head and covering your whole body until the water is all gone. When you are done, bundle up and keep warm so not to get a chill and sit silently for a few minutes. When you are all done, you give the plants back to the earth as a form of thanksgiving. We always say thanks to the earth for all the good things she gives us and we lay the used plant materials respectfully back on the earth. Eighty-five to ninety percent of herbs are for women's healing. For that we are so grateful.

What you take, you must give it back. It's not all about taking and taking. You must give back again! It's all about giving and receiving. We should work this way with humans also. Humans need friendship. What I have, I give to you and what you have you give to me. We share. It's not about give, give, give to me, me, me. That just isn't right.

Some people don't know this, but if you go to a Maya monument and you see a little stone, you should not just go and take it. Unless you have permission, do not take even a single stone.

Maya Medicine has been practiced since way back. My ancestors in the rainforest detected and treated spiritual ailments in the ways I am telling you and probably with good success because they lived to an old age. All my ancestors practiced what we are talking about right here.

They had Maya blessings and prayers, herbal baths, teas, the egg clearings and the ceremonies.

Maya people are very careful about eclipses. The energy during an eclipse changes your aura and energy for two weeks. Out of nowhere, a close friend may say something they would normally never say. The eclipse gives you a bad influence to speak something that you shouldn't. During an eclipse is a good time to stay home and say your prayers. Pregnant mamas are especially vulnerable to the energy from eclipses. We believe that the baby could get a cleft palate or could be miscarried if the mama is directly exposed to an eclipse.

Way back, in the Maya villages, there would be a shaman, an acupuncturist, a bonesetter, a snake doctor, one who collects the herbs and helps people do the plant spirit baths. Sometimes one person would do all those jobs. That is a big pile of work! This person was well respected. My father said that way back then money was very scarce so they did trade for their work. Your patient might give you rice, corn, beans, fruit, a chicken or do farm chores to pay back what was given to himself or his child. It was a fair trade for receiving healing and an important way of showing gratitude.

In life, we can't just do work, work, work! We need to have fun in our lives to keep us happy. We need to go by the sea, look at the waves, walk in the sun, pick wildflowers and do things that really please us. What I mentioned are just some of the things that could make us happy. Only you know what those things are for you. Keep laughing! This life is too short. We never know what is going to happen tomorrow, so because of that, I tell people to live one day at a time and try to be happy. It's not always easy but I think if we start our day by saying, "Today, no matter what, I want to be happy!" then you can overcome anything. Some of the things that have happened to me, oh boy, if I told you, you would never believe it! Envy over me is very big. I get attacked from all corners. If I feel I am getting attacked, I pray Psalm ninety-one and also Psalm twenty-three. I stick to them every day. My house is not belonging to me. It belongs to my kids, but mean people in the village all think it's my house, that I own it, so I must be rich. That is why they envy me here in the village. I am envied also because I have friends and students coming from the U.S., bringing me presents. All of that makes them jealous and they send me bad vibes. I just keep saying my prayers, burning my black copal and doing my plant spirit baths. You never do know what is going to happen, so we need to say our prayers every day and appreciate what we have.

As a healer I know my limits. I'm not going to take a knife and cut you open. That's the doctor's business. Healers have deities, plant allies and spiritual guidance, so no healer works by herself. A healer should not boast and say, "I know it all and I did it all!" That is making themselves way too big! We do not heal people. It is only God that heals, not us.

Miss Beatrice's Healing Hut, Santa Familia, Belize (photo by Katherine Silva)

Maya Plant Ally Number Four: Basil

Ocimum basilicum

Basil leaf with Marigold and Rosemary in Miss Beatrice's Hands

(Photo by Monica Jochum)

Basil is a very important plant. It grows wild here in Belize and we also cultivate it. We call it *albahaca.* We use it for our spiritual baths, most often with Marigold and rue. We pass a basil leaf over a hot flame and drip the warm juice in ears for ear ache. A womb steam bath is made from basil leaves and given to women before menses or after childbirth especially in cases of a swollen uterus or ovaries. Also in the case of a swollen uterus, we give a lady a tea made with three leaves of basil crushed in a cup with hot water poured over it and steeped for ten minutes. We recommend she drinks the basil tea for nine days in a row. Basil disinfects, cleanses and blesses all at the same time. Basil is a plant deity that I work with often when I say my prayers into people's pulses. We Mayas use basil in our house blessings by soaking three sprigs of basil on a Friday in a crystal bowl under the full moon. We then make three crosses on the ground near the front door with a sprig of basil and the full moon basil water. Basil is an important ally to always have nearby.

Chapter Five: Our Sacred Blood: Maya Medicine for Women

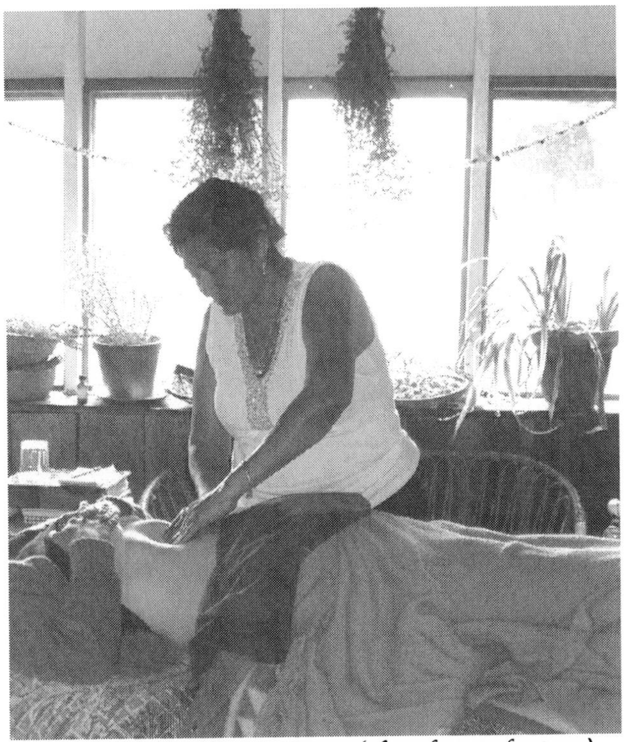

Miss Beatrice giving a treatment (photo by Sandra Lory)

"Trees can put down roots in a stony place and thrive. In this way, so can a woman birth a baby. She has to be strong. Nobody else can do it for her. She just has to do it herself!"

Before we begin talking about women's health and women's ailments, there is something I will say which is very important. We Mayas have a goddess whom we honor very much named *Ix Chel* (pronounced eesh chel.) She is the maiden, the mother and the crone. She is the great medicine woman. She is the one who can be called on for women's ailments, fertility, childbirth and anything pertaining to women's health. The maiden Ix Chel is more about weaving and child-birth. The Crone is more about healing and wisdom and the moon. We Maya honor Ix Chel when we are birthing by lighting a little white candle, offering white roses and saying prayers. Ix Chel is the moon goddess, the medicine goddess, the water goddess and the rainbow goddess. She brings us healing and awareness. She is the one to always have nearby when addressing women's concerns.

Many women love to plant gardens and work with plants. When we Maya women want to experience the magic of Ix Chel there is a little something we do in our gardens. If we plant our flower garden when a rainbow is in the sky, the zinnias will grow in all the colors of the rainbow instead of just one color. That is only a little bit of the magic of Ix Chel, the rainbow goddess.

Ix Chel is a very powerful deity and she is always there to help whosoever calls on her with faith and reverence.

Ix Chel is the consort of *Chac* the great rain god. When Ix Chel gets angry, she sends floods. Rainbows are also messengers. Depending on if they are occurring in the East, the North, the West or the South the message is different. Sometimes they are even found in a little pool of water. A rainbow in the west denotes war, famine or other hard times. We Maya pay attention to the signs of nature because they are messages from the Gods.

Ix Chel in one of her many forms, Cayo district, Belize (photo by Janet Caspers)

I am now going to address some ways that Maya women keep their physical bodies healthy through all the stages of life. I will also talk about some female problems that Maya medicine can be very helpful in resolving. In Maya medicine, even when the problem is physical, we always address the mind, spirit and emotions of the person as well. It took me many years and a lot of sacrifice for me to learn the things, I am telling you. If you want to learn how to do these things you should train with someone qualified to teach you and practice a lot. But don't worry. Many of the things I am sharing here are simple things that you can do in your own daily life that will help you a lot if you choose to apply them. Also, you could go to a Maya healer to receive the more advanced treatments and perhaps learn how to do them as well.

PREGNANCY

In the Maya way of thinking, you should not have two babies close together. My grandmother always said it is best to wait until two years time has passed before getting pregnant again. If you got pregnant before two years time, she would say that you weren't taking care of yourself. If you have two babies too close together, your uterine ligaments will become overstretched and give you trouble later. Without time for healing in between pregnancies, a prolapse of your pelvic organs is likely to happen. Also there is not enough nourishment in your blood for yourself or your new baby when the babies are spaced too closely together.

During all my pregnancies, my grandmother massaged my belly from the third month onwards with her special womb massage. She did this for all her pregnant mamas. She always took a case history of the pregnant woman so she knew if it was safe to do the massage or not. She also stayed close to her intuition and that of the pregnant mama. If the woman chronically miscarried, she did not massage her belly until she knew the pregnancy was secure. After the first trimester, she would lift the uterus up so it's not hanging down so low. That is better for the mama and the baby. They have more room and better blood flow. I do all of these things for my pregnant clients as well, just as my grandmother taught me to do.

We never give a deep foot massage to pregnant ladies because it could cause a miscarriage.

My grandmother said, "Zero deep foot massage in pregnancy is what is needed."

Only in labor, deep foot massage is good for pregnant women.

Thanks to my grandmother who never went to school, how she knew this I don't know, but she said, "Do you know what will be caused by doing deep foot massage on pregnant women? You can cause the woman to have a miscarriage because everything is connected."

For that reason, we don't do deep foot massage on pregnant women. If a pregnant lady comes to me, I just do the prayers and bypass the teas and womb massage unless I am sure everything is okay with her. If I know she is passed her third month and the pregnancy is healthy, I give her womb massage and teas with her prayers. I avoid deep foot massage, especially in the ankle, big toe, little toe or arch of the foot. I just focus my massage on the side of the body and the abdomen. I check if the baby and womb are centered. Sometimes the baby gets stuck only on one side and that is a big problem. With her baby way over on one side, the lady may complain of sciatica. The uterus is out of position and of course it is pulling on everything and blocking the energy causing her pain. The uterus is sitting in the way of the flow, blocking the energy and movement in the lady's abdomen. This causes back pain, cold feet and other discomforts. The uterus is sitting way over on one side or the other because, when she got pregnant, she got so cozy and just slept on that one side and the baby got cozy right there on the one side as well and just likes it right there. With her uterus way over on one side she could get a backache and her labor may be difficult. Centering the womb will help relieve sciatica and make her labor much easier. Women should alternate sides that they sleep on during pregnancy so that the baby doesn't get too cozy on only one side. The mama can prop a pillow under her growing belly whenever she is lying on her side to keep the womb from sagging over too far.

Most of women's health issues are caused by a wandering womb. If the baby is way over on one side, the mother could have a very difficult labor. The baby could even die because he tries to come out, but instead of coming out easy, his head just keeps banging into the side of the mama's bony pelvis. If you just center the womb, the baby flies out. Without the womb massage, there are a lot of C-sections. With the womb massage, the baby is in the right passage, the right canal, and he comes flying out. That way both the mama and baby are happy.

We like to keep it simple. We just massage the uterus back to the center. The pregnancy massage is different from the womb massage for women who are not pregnant. We are very gentle. We know what we are doing and we use a light touch. We then lay the pregnant woman on her side and gently push down on her side all the way up and down her body which helps baby and mama to relax. It's just like a little pampering. Centering the uterus is still the most important. We can do prayers and baths and teas for pregnant mamas as well and that will help a lot.

The *faja* is a cloth which we wrap around our bellies to support the uterus. The faja can be worn the first three or four months of pregnancy, but after that, it is too tight. It could cause contractions if worn tightly past the fourth month. After the baby is born, the mama can wear the faja again for three to six months. She will know when her body is healed. When her body is done and her uterus is strong and right back up where it belongs, she can take the faja off.

We have to make a good home in the womb for that baby. It is best if the mama has belly massage and teas before she gets pregnant and then gets the massage at least monthly during the second and third trimesters of her pregnancy and then postpartum as well. If the baby is in a good environment, she will just want to grow and grow.

Pregnant mamas should eat lots of *kalaloo* (wild amaranth) and other greens like chaya, chard and wild spinach. Pregnant mamas need beets, carrots, chives and lots of protein like fish, chicken and eggs. The pregnant mama should not eat ice, bathe in cold water or wear tight things.

My grandmother said to all the pregnant mamas, "Not too tight, not too cold."

If the baby is really low too early in the pregnancy and the mama is having contractions too soon, I ask her if she is drinking ice and bathing in cold water. I get her drinking warm things and taking warmer baths and I lift the uterus and give her some teas and then everything is okay.

In late pregnancy, I encourage the mama to prop her feet up on a stool or lay down. Sitting too long without her legs propped up can cause the uterus to be pulled down too far, and this can lead to preterm labor. During early and mid pregnancy, we Maya midwives center and lift the uterus to keep it in the best position possible, but during the last month of pregnancy, we only center it, and we don't lift. We leave the baby down where it is waiting and ready.

I tell my pregnant mamas that walking is the perfect exercise for pregnancy. Sitting too much during pregnancy is not good for the mama's circulation or the position of the baby. Too much sitting makes the uterus low and the mama will have to pee all the time.

Maya midwives don't go poking their fingers and other things in vaginas as long as they can avoid it. My grandmother said that putting anything unnecessary in your vagina is like putting dirt in your eye. Just think about what dirt in your eye does to you!

In pregnancy, it makes a difference what the mama focuses on in her mind and heart. During Abimael's pregnancy, I focused on long curly eyelashes and he definitely got those!

Pregnant women can do their own prayers. In pregnancy, it is best to avoid heated arguments or being in places that are stressful. When the mama gets in the mood to eat something particular, she better get what she is craving and eat it because it is the baby who is wanting it and needing it! I tell my pregnant clients, "If some food sticks in your mind when you are pregnant, well, go get it and eat it."

When people say, "She has passed her due date," we Maya believe that is not always true. We think that if a woman conceives on the sixth of January, she is probably going to give birth about the sixth of October, if she's already had her first baby. But, if it is her first baby, it usually goes about nine months and seven days from conception. If the seventh day passes and no baby has been born, we watch that baby and mama very carefully. But we like to say that each baby knows the exact right time to come.

If you look up at the full moon, you will see a mother breast feeding her child. The moon is very connected to women and our cycles, our pregnancies and our births. Usually we have an idea when we got pregnant. We check the moon and we know our cycles. How can somebody tell you that you have two more months to go when you know it's already time?

It happened the other way around with me when I was pregnant with my daughter Berta. She was the one who went breech. I had a period each month during my first two months of her pregnancy, so I didn't think I was pregnant.

On the third and fourth month, my stomach was getting much bigger and my clothes were tight, so I thought I was getting fat, and then in the fifth month, I felt some movement and I said to myself, "Oh my goodness, I am pregnant!"

So, I started to count from the third month of my pregnancy thinking it was the first month. Then at what I thought was the seventh month, I got a really bad contraction, and I suddenly wanted to eat grapefruit *really badly!* I could see the grapefruit all ripe and delicious in my neighbor's yard. My daughter went to get some grapefruit for me, but my neighbor told her that they were not ready. I could see out my window that they were very yellow. I went to sleep for a while thinking about that grapefruit I wanted so much. Then I did my washing. I did half my washing but I couldn't finish because of the contractions!

More and more contractions came so I said to my husband, "This baby is going to come out soon and you need to do something because, I'm having this baby!" I thought I was just seven months and so I was a bit concerned.

I went to look for a midwife. When she came, she massaged me and said, "No, you are not only seven months. You are full term and the baby will be born in a few hours. And by the way, the baby is breech!"

Oh my goodness, I was shocked. She did hot compresses, she warmed me with oil and she gave me chamomile tea to drink. The pain got worse. In birthing, sometimes we have to work hard. My other labors were easy, but this one was breech. I hadn't gotten any massage from my grandmother, and oh boy, could I feel the difference. At the midwife's suggestion, I drank my husband's urine and then the baby was still not coming out. We then put my husband's cap on my abdomen and said some prayers, but still the baby was not coming out.

The midwife went to get some ants from the trumpet tree and made a tea and said, "Right now we are putting you to the test. If this baby is your husband's baby, this baby will be born soon."

In twenty minutes, I was pushing her out. The baby looked more like her father than any of my other children.

I know all the things the midwife did really worked well. If I were having a breech baby today, without a Maya midwife, the doctors would have cut her out! If I had a massage a week before the birth, the midwife could have easily turned the baby, but since labor was already in progress, it was too late. Breech birth is not easy.

All my other births were easy because I was massaged by my grandmother throughout my pregnancies, so my advice to women is, "Get your belly massage! That's what I have to say!" I now know the difference between having the massage and not having the massage!

Maya midwives also attend women having miscarriages. However, if the miscarriage is complicated, we take that woman to the doctor. Miscarriage can be very risky. A woman could bleed to death very quickly. I don't take charge of those kinds of things. I take them to the doctor. On many occasions, women come to me with only a threatened miscarriage, and I can make them the tea. We have many teas that we use for each circumstance, but hibiscus tea is the one that coats the uterus, builds the blood and stops the bleeding. I also do a special treatment of wrapping the woman with a long cloth if it's the first trimester. Further along in pregnancy, the wrapping of the cloth technique up and down the body is too tight for the baby. For threatened miscarriage, I put the mama on total bed rest. If the baby is too low, I put her feet up. The tea I give for a threatened miscarriage is red roses or hibiscus flowers. With prayers, I use five opened and five closed and nine leaves boiled in a quart of water. I collect the plants with prayers and say prayers for that woman and the baby. The tea is served cool and just sipped on while staying in bed. Cold contracts and heat expands.

When I make a tea for a lady in labor, I give it to her hot or warm, but not when there is a threatened miscarriage. That's when I give it to her cool. If the baby is further along, I give the tea and some prayers, but no wrapping with the cloth. I don't give her any massage, especially not foot massage. I just put her on total bed rest.

There are many teas I really like such as cinnamon tea and red china root. All the teas that we use for threatened miscarriage are blood builders. If blood is gushing out below, the blood builders are building it back up above. Red china root is a blood builder. We know this because it is red. Wild yam is white so that would not be the one to use. If the fetus is dead, but it is stuck inside, we give garlic, Moses in the cradle and thin oregano (oregano delgado). It is cultivated. The thin oregano is used for problems like fibroids. The fat oregano is good for earache. All those things have a hot energy to flush things out. To flush out a dead fetus, heat the belly with a hot compress and give the mama a dose of castor oil as well.

I think that women can have babies when they are older, but it may be harder on their bodies and harder for their joints to open in labor. I had my last one at thirty-nine, but I have seen people have babies even in their fifties after getting some womb massage.

Pregnant women benefit from drinking hibiscus tea all throughout pregnancy to nourish themselves and their blood. Hibiscus is rich in iron. Red china root and red rose petal tea are both good sources of iron for pregnant women. Eating a lot of greens like chaya, spinach and amaranth as well as protein such as eggs, chicken and fish is also very important. From the end of the seventh month of pregnancy onwards, we give the mama slimy things like okra and *nopalitos* three times a day. The slimy foods help the passage to be smooth and then the baby can fly out fast, when it is time.

We feed pregnant mamas a lot of spinach in her last month or so of pregnancy. The spinach is to nourish her and to help her perineum to stretch without tearing.

We encourage pregnant women to stay happy and relaxed, to focus on positive things, and of course, to get womb massages often. Near the very end of the pregnancy, I recommend that the mama begin gathering rain water, which will be used for the baby's first bath.

BIRTH

All my labors, except the one in which I got no massage from my grandmother, were about an hour or two and not hard at all. My grandmother was my midwife and she knew all the right things to do.

In a Maya birth, we use hot water compresses and hot water bottles on the mama's belly, special teas, massage and more to help the laboring mama have an easy birth. In early labor we give a tea of chamomile with a clove of garlic served hot. We encourage the mama to be upright and walk during labor. We oil and massage her perineum during the labor so she will never tear. We make sure that when she is pushing the baby out, she has her heels pushing on something hard like the wall or a bed board. We give the mama a rope to pull from either side of the bed or from above, and we put a cloth in her mouth, if she is okay with that, so her scream goes downwards and helps push the baby out.

We always say that if the birth happens on a new moon, the birth will be slow, but if it is on a full moon, it will be fast. Babies born on the full moon are generally tough guys or girls. Babies born on the new moon are sensitive and loving. Women usually deliver during the same kind of moon in which they conceived whether full, new or in between.

Sometimes it is a false labor. That is okay. Those pains are not really false. Those pains are like a little chick pecking its way out of the shell, peck, peck, peck. Every little peck brings the chick closer to the light of day.

One trick we Maya have is that we can predict the time of the birth with a lit candle and we can also determine if the labor is a false one or the real thing with our little flame.

While the mama goes through her labor, we put hot compresses or a hot water bottle on the mama's belly and we give a warm olive oil massage on her belly in big circular strokes to keep the contractions moving along. Warmth expands everything and helps the woman's body to open up and let the baby come down. We also give deep foot massage during labor.

During the whole labor we wear sterile gloves, and we warm and massage the perineum with warm virgin olive oil. When the time is coming closer for the baby to be born, we stretch the perineum with gloves that have been dipped in warm virgin olive oil. When there is some mucus with a little bit of blood coming out, then we know everything is moving along nicely and soon it will be time to push the baby out. For a special blessing and for protection, Maya midwives smudge the birthing room with dried rosemary right before the baby is born.

A lot of times after the mama passes the little bit of blood and mucous, she wants to poop or begins vomiting. This helps with the birthing process because then there is more room for the baby to pass through. Sometimes we tickle the mother with oil and a turkey feather to make her vomit, which also helps her dilate faster.

Sometimes we take a little piece of bay cedar bark, like a four by four or a six by six section. We pound it and soak it in water. We give that water to the birthing mama to drink and the baby comes out quick with no problem, not even one.

Sometimes if we are not sure who the father is, and if the mama is willing, we put the probable father's hat on her abdomen during a long labor. If the baby comes out in a timely way, then we believe it is that man's child. If it takes a very long time, and there is a lot more pain after the hat is there, then we think that baby may have another father!

If the baby is big and slow to come out, we have a little trick. We put heated olive oil on our gloves and put our fingers under the baby's arm pits and wiggle him out. He will scrunch up a little bit inside the mama and wiggle himself out.

We also squeeze the baby's shoulder in a little bit as it passes over the perineum because usually it's the big wide shoulder that tears the lady not the head. This way we can keep her from tearing.

If the placenta is slow to come out, we give hot allspice tea and a womb massage. Another special placenta tea is given to the mama to help ease out everything from the womb and make a good flush. This tea is made from Moses in the cradle, garlic, chamomile and rosemary, in equal parts, boiled and given to the mama. It should not be served hot, but just kind of warm. This tea helps expel everything.

If the birthing mama begins hemorrhaging, we give her a cool tea of nine hibiscus flowers and a three inch stick of cinnamon in a quart of water. If the bleeding is too serious, we take her to the doctor because too much bleeding is no joke. Birthing a baby is not an easy thing to do, but women are strong and we can do it!

I tell my birthing mamas that trees can put down roots in a stony place and thrive. In this way, so can a woman birth a baby. She has to be strong. Nobody else can do it for her. She just has to do it herself!

BABIES and POSTPARTUM

In the Yucateca Maya tradition, a baby's first bath is made out of rain water and flowers. The first bath is given only on the eighth day. We believe that it's a bad omen to bathe a baby on the seventh day, because if we do, that baby may get sick and die.

When the baby is first born, we just wipe the baby with olive oil all over his little body. That is all he needs. The oil is warmed in a little container and put on a cotton ball or a sterilized washcloth before it is rubbed on the baby's skin. For us, olive oil is anointing. The baby is cleansed with olive oil when he or she is born up until the eighth day.

On the eighth day, we gather red hibiscus flowers, red roses, white roses, pink roses, gumbolimbo leaves and blue vervain, and we boil them in rain water. Before the baby is born we collect rain water to make this first bath. We boil the plants in the rain water for five minutes and set it to cool, and when it is tepid, we give the baby his first bath. First we stick cotton in his ears and a little oil in his navel, because sometimes the navel is not totally healed and we don't want water getting in there. Then we put the baby in the bath. After his first bath, every other day we give the baby a bath with the rain water and flowers. The flower bath protects him from wind invasion, from evil eye, from the sun, from skin conditions and all little things like that. A baby's bath is not poured over him. The baby is put to soak in the rain water and flowers and always bundled up and kept warm afterwards.

Here in Belize, we use the flowers that we have abundantly growing in the forest. Where you live, you can use some gentle nontoxic flowers that have not been sprayed with pesticides for the baby's first bath. You can use chamomile, catnip, calendula, basil, roses, *malva* (mallow) or some very gentle plants you have growing where you live. Remember, nothing too strong is to touch the baby.

Sometimes we put a mixture of balsam copaiba, castor oil and olive oil in equal parts on the umbilical cord. We mix all three ingredients together to treat the umbilical cord. If treated properly, the cord falls off in three days. Even after it has fallen, we put a little oil on the belly button before bathing the baby, so water doesn't get in there and create infection. We do little things like that just to make sure.

Sometimes babies born in the hospital take a whole month for their cord to fall off. In that case, we light a candle and warm the oil with it.

We put the warm oil all around the cord and in three days the umbilical cord dries and falls off.

We tell mamas, "Watch out if that cord falls really quick because that is a sign that if you do not protect yourself, you could get pregnant again just like that! Then you will have two babies back to back!"

We don't take our babies out until after the umbilical cord is totally healed or even longer. We kind of hide our little babies away with us as a way of protecting them. Also, in the very beginning is a good time for the mama to take it easy; she doesn't need to be out on the town.

If the umbilical cord is bleeding, but not infected, we put a nutmeg pod in a little sterile pillow and tie it on top of the cord and change it every two days. That clears up the bleeding and dries out the cord so it can heal properly. If the umbilical cord is red around it, has pus oozing from it or is inflamed, it is infected. That is when we take the baby to the doctor because that is no joke.

We advise mamas to protect the baby and to avoid taking him out and about until he is older. When the baby is old enough to go out in the public eye, a mother can put a little red coral bracelet on his left wrist for protection.

We Mayas do not leave our babies alone in another room or put them in car seats or strollers. We don't really have those types of things. We like to nurse our babies a lot and keep them close to us while we are awake and while we sleep. When we stay together, both the mama and the baby are happy.

Wherever the mama and baby are sleeping, we make little angels and crosses out of white wool and sticks. The angels have a cross made from sticks on their backs and are wrapped with white wool. We also burn rosemary in the room where the mother and baby sleep. We do all this to protect them and bring good, healing energy.

If the mother is a healer, she should do no healings until her baby is one year of age. Otherwise any energy she picks up could affect the baby.

How a mother is treated postpartum and how well she takes care of herself right after her baby is born will affect her health for her whole life. Three days after the baby is born, it is time for the mother to take her *bajo* (womb steam bath), her plant spirit bath, get *la sobada,* (a womb massage), and receive the wrapping massage with a *faja* (long narrow cloth).

The womb steam bath is made from astringent herbs like yarrow, rue, basil, artemesia, rosemary, guava leaf, uva ursi or red roses and some soothing, uplifting and calming herbs such as blue vervain, Saint John's wort, peppermint or chamomile boiled in water for a few minutes. We make a combination of those herbs or sometimes we just use basil. When we make the steam bath, we save half a glass of that strong, warm herbal water for the mama to drink. Then, we put that big pot of boiled, steaming water with herbs beneath her slatted chair. She sits on the steam wrapped up in blankets to hold the heat in. She has her *chonies* (underwear) off, and the blanket is wrapped all the way around her. In this way, we make a tent of steam around her naked lower half. We let the steam rise up beneath the blanket for twenty minutes. This helps to flush and tone up the uterus. The herbal steam rises into her womb and helps clean out anything that still needs to come out. The steam cleanses and honors the womb. The astringency of the herbs helps the uterus to contract and return to normal size. The soothing and calming herbs relax the mama's nerves and help heal her perineum. We make sure the steam does not burn her. If it is too hot on her thighs, we lay a washcloth on them. If the steam still burns her skin, we leave it to cool a bit longer before starting.

After her womb steam, we wrap the mama up and keep her warm. We never leave her to be exposed to a cold draft after a warm treatment. We have her rest in a warm, comfortable place for at least ten minutes afterwards so that she does not get a chill or put her feet on a cold floor.

We give a postpartum woman a plant spirit bath with boiled herbs rather than mish mashed raw herbs, just to keep everything sterile. A postpartum woman's pores are wide open, so we cannot risk infection by using non-boiled water and uncooked herbs. We also do not put a postpartum woman in cold water. The bath must be very warm and poured over her in the warm sunshine or in a warm bathroom. We wrap her in towels and blankets afterwards to protect her from cold drafts. The bath is made with prayers, red hibiscus flowers, blue vervain, neroli, peppermint, basil, rosemary, roses or whatever sweet, organic, harmless plants we have growing nearby.

Postpartum women need *La Sobada,* (womb massage). La Sobada is always done by a midwife trained well to perform it. The massage lifts and centers the uterus so that as it returns to its pre-pregnancy size, it will be in optimal position. When the mama returns to her monthly cycle, she will be regular and pain free.

The faja cloth wrapping treatment is quite simple; and anyone can do it. Two people are on either side of the postpartum mama. The mama is lying down and resting. The helpers wrap a *faja* or *rebozo,* which is a long cotton cloth, about five feet long and two feet wide, around the mama's forehead. They cross the cloth over her head and then pull the cloth from either side. This gives the mama a feeling of containment as if she were being put back together after being wide open since giving birth. Then the cloth is moved down to the shoulders and the same wrapping and pulling happens. The breasts should be left out of it and not squeezed at all, as they are full of milk and quite tender. They move down the abdomen, legs and feet until she has been wrapped and squeezed at each point. She will get a calm feeling from this treatment as if she has come back into her body and back into herself.

I make my new mamas a tea, made from one tablespoon chamomile, two tablespoons of rosemary, one glove of garlic and one tablespoon of anise seed in a quart of water. This tea helps everything to flush out. The mama may see clots pass. After one month when everything is flushed out, she can stop the tea. I give the mama *la sobada* on the third, fifth, seventh, and ninth day postpartum. Normally I avoid womb massage when a woman is bleeding, but postpartum bleeding is different from menses, so it is okay. Of course, remember, it must be done very gently by someone with experience.

Both the steam bath and breast feeding help to shrink the uterus. The massage helps the uterus settle into a good position. When a woman is treated with this special pampering, she will recover very quickly, regain her strength and stay healthy.

After the mother has given birth, we offer her a special diet. She could eat lean soup such as chicken or fish soup. That will be good for her. It is best for her not to eat anything that is pickled, cold or sour because that can cause the blood to clot, and then it won't gush out freely. Also it is best for her to avoid oily foods because that will slow the healing inside of her. If the mama has onion in her soup, she and the baby are probably going to have colic. If she wants to bypass that trouble, she shouldn't eat onion in her soup for the first month after the baby comes.

Sometimes on the third day, we give the mom castor oil to flush everything out. We Mayas love our castor plant and we use castor oil for many things.

The mama will usually bleed for two weeks. Then, for another two weeks she is just spotting and spotting, then only pink and white is coming out. Then she is all done and she can say, "I am free and I am okay now!" From that point on, however, she could get pregnant again so she must be very careful. If she totally breastfeeds all around the clock, it's less likely she'd get pregnant, but she definitely could. So she has to watch out and be very careful.

If the baby has colic, I take the skin of a garlic clove and one marigold flower (yellow one) with nine little leaves. I boil it for five minutes in a pint of water and set it to cool. In North America, you may have a different kind of marigold and maybe you don't know if you can trust it, so you could steep some chamomile flowers instead. Whatever you do, it must be organic and sprayed with no pesticides. We sweeten our baby's tea with wild honey. I know in America everybody is afraid of honey for babies, but here we have the wild honey that causes no problem for babies. In fact, it is a medicine that keeps them healthy. Be careful about introducing anything besides the mama's breast into the baby's mouth. We put the tea on a little cotton ball for him to suck. Introducing all kinds of things into the baby's mouth can give him mouth thrush and confuse him as well. If one of our babies gets thrush, we wipe the baby's mouth with her own urine or with wild honey. It works really well.

If the baby has cradle cap, we rub his head with olive oil one half hour before he bathes, and then with a fine comb gently scrape. We have the John Charles plant and we boil it and that's what we wash the baby's head with. In America, you can substitute something like artemesia, yarrow leaf and flower or even basil.

Sometimes the mama's milk doesn't come. This can happen because maybe she doesn't want to breast feed or she thinks her breasts are going to sag. Her breasts are going to sag anyway, so she might as well put them to good use. In case the milk is not flowing, we boil a little bit of chamomile and rosemary and show her how to wash her breasts with it while it's really warm. If the mama needs help bringing down the milk, we make her some porridge out of ground pumpkin seeds. Sometimes on the third day, the mama gets a big fever when the milk starts flowing. Sometimes, with first time mamas, her nipples crack a little bit, but she still has to make the sacrifice and give it to her baby. We put virgin olive oil on cracked nipples. Olive oil is at no time bad for babies. It will soothe your breasts, and if you keep at it, everything will be okay.

If the baby gets used to using the bottle, he is going to reject the breast because with the breast he has to suck and pull really hard and the bottle is easy. Milk from the bottle just pours out so, the baby gets lazy and will make no more effort to pull and suck at the breast. As a breast feeding counselor, I tell mamas to give the baby the breast first because the colostrum is the baby's first antibody. When you breast feed, you are giving that baby your antibodies. This is especially important for the first three months. It continues to be the best food for the baby for a good long while. If you miss giving that antibody because you didn't give the breast to the baby, then you can have tons of trouble like diarrhea and colds. Without breast feeding, you could end up living in the hospital.

If the baby catches cold, take one half teaspoon of almond oil and mix it with your own breast milk and give it to the baby to drink slowly by slowly. Also we heat almond oil and put it on the baby's soft spot, his palms and the soles of his feet and then put little socks on him. We put a little cap on the baby. Babies with colds should not be exposed to drafts. They should be kept warm. Keep this treatment up for a whole week or so. Sometimes this oil treatment makes the baby pass or vomit some mucus and that is good. You might see big pieces of mucus. That is the cold coming out.

If the breasts get engorged, we get the fresh castor bean leaf and put on Vicks vapor rub and a pinch of salt on the big leaf. Then we put it on the breast avoiding the nipples themselves. In America, there are castor plants. I think there are purple ones. I saw them in California. That California castor bean leaf should work the same as ours. I never saw the green ones there that we have here but the purple ones should work fine. Maybe people in the U.S. could plant the green ones. Then they could use castor leaf for many things.

When a baby has colic, he will cry and cry so much! When I put my finger on his tummy and tap it, it feels like a drum so I know he has colic. So I give him his first bit of tea and he belches and farts. For the tea, I use chamomile with garlic peel. Chamomile is really good. Our Belizean marigold flower with garlic peel also does the trick. Garlic is too strong for newborn babies. That's why we use only the skin. Other herbs we use to make tea are fennel, dill or anise water, but get ready, because the baby is going to fart. With babies we have to be careful, they are very vulnerable so we don't use anything strong.

When the baby gets older, like four or five years old, and has a belly ache, we put four pieces of garlic skin in fennel, anise or chamomile tea and give a good foot massage with olive oil.

For ear aches, we put basil leaf over a hot flame and squeeze the juice in the ear. If we use the wild basil for an earache, we have to get a whole lot, because it's not so juicy, but the cultivated one is very juicy; just three leaves will do.

The gourd bowl flower and wild cucumber flower both look like an ear. They work for earache as well. This is the doctrine of signatures. Fat oregano can work, too. We always treat both ears, even if only one hurts.

After three months, when the baby starts biting the nipple, or if she is using her finger to press on her gums and kind of biting everything, we roast a garlic clove and rub it on the baby's gum. This helps the tooth to burst forth. There is sometimes vomiting and fever with teething. We still give the marigold or chamomile tea with garlic peel. Something special is happening in a baby's body when she's pushing out a tooth. If she's already eight months, we put a little garlic mashed into a tea to make the diarrhea and vomiting stop. If the vomiting and diarrhea continues, we give warm cinnamon tea.

For fever, we put a poultice of life everlasting leaves, gumbolimbo leaves, castor bean leaf, *nopalito* pads or aloe vera plant on the baby's forehead. That really works to pull out the heat. Babies are a lot of work from start to finish but they are well worth all the trouble.

MENARCHE

Way back, everybody started menses at age sixteen or seventeen, and now it's starting from age nine or ten! I heard that this is because there are hormones in the chickens and pesticides in the plants. That's why I stick with my own chickens. I raise chickens, keep a garden and gather my wild *kalaloo* (wild amaranth greens) and wild spinach. Wild food is the best. If a young girl receives her abdominal massage, her period will start nice and easy and get regular fairly quickly.

THE MENSTRUAL CYCLE

My mother taught me that women are not to eat lime, lemonade or cold or pickled things during menses because it clots the blood and then everything won't flush out properly. If everything doesn't flush out properly, there will be stagnant blood left behind and stagnant blood leads to trouble. Women should not put their feet on cold floors or cold ground because the cold travels right up their legs and gives them clots and cramping! Nature intended menses as a way for the body to cleanse itself. I hear that nowadays, for up to a whole year or two, some women don't have a period. No wonder they have fibroids! What will be the consequence of taking a pill that stops women from having periods? Stagnant blood! Imagine, if for a whole year, I didn't put out my garbage. Say I just left it by the door. What would be the end result of that? I would have to climb out from my window. Then the garbage would cover that window so I would have to go out another window. Then that window is covered so where am I going to go? I will be trapped inside. That's the way I think about that pill. It doesn't happen overnight, but little by little, pieces of old blood stay inside. After a while, the stagnant blood pools and builds up in the uterus. We believe that if you eat a lot of cheese and milk it becomes even worse. This is what leads to fibroids and many other problems.

My grandmother taught me about an herbal formula that is very handy. This blend of tinctured herbs is something that every woman would find useful. It is an herbal formula my mother and my grandmother used from way back. It contains rainforest herbs that balance the hormones, increase warmth and circulation to the uterus and flush out the uterus. We only give it when the woman is not bleeding, because it could make her bleed too much if taken during the period. It will regulate her cycle and flush out any old blood during each menses. I recommend that she take it for two weeks before her period. She needs to take the formula for at least three cycles and combine it with la sobada, the womb massage, to get her improvement.

Pads are the best thing to use to catch our blood. When we use tampons, or those diva cups, the blood cannot flow out properly and that can cause problems.

If you get your womb massage, you won't have much cramping. When women have light cramps at the beginning of menses, ginger or allspice tea can be used for relief.

However, if you are bleeding too much, you shouldn't take ginger tea because it will make you bleed more. I suggest chamomile instead, which is mild and won't make you bleed more. It will just wash out what's not supposed to be there and ease the cramps. Right before menses, we do our *bajos*, (our womb steam baths), with marigold, basil and rosemary. We feel it is important to do these simple things; massage, herbs and steams to promote a healthy flow.

To make a *bajo*, we collect our plants with prayers. We choose basil, marigold, rue, artemesia, sage, St Johns wort, red roses or other plants that are astringent and antiseptic. We mish mash our herbs with prayers in a big pot of water and boil it until it's steamy. During the last week before our menses, we sit naked from the waist down on a birthing stool. We wrap up in a thick blanket. We make sure our feet are warm, and no draft is coming into the room. We let the steam rise up through our vagina into our womb. We burn copal and say prayers just resting there for twenty minutes. Afterwards, to make sure we catch no chill, we go to our bed and rest for at least fifteen minutes. Then we are done. That helps flush out everything not needed from the womb, such as old blood and also negative energy from certain men, abuse or trauma.

Young women, during menstruation, need to take care when riding bicycles and horses. What could happen when bumping on bicycles and horses? The uterus can get tipped. Also, falling downstairs or from a horse, or any big fall that happens to us as a young girl, can affect our quality of life if our uterus gets out of position. The trouble does not happen overnight, but it's going to happen. If your uterus is tipped, the first time you have your menses, you will start with intense cramps. You may also have constipation, headaches, lower back pain, weakness, vomiting and even inability to stand. With any of these symptoms, the uterus needs to be repositioned. All you need is la sobada, a good womb massage.

The person giving you your massage can check the position of your uterus. If it is too low, it gives you "tee hee pee pee," which means you pee whenever you laugh. Sometimes this problem could be a kidney problem, but normally it's a low uterus sitting on your bladder, and only womb massage will help. You could drink all kinds of herbs and take all kinds of pills but you are only treating the symptom and not the cause. Womb massage treats both the symptom and the cause together. If the uterus is tilted back on your colon, that takes a good time to fix with lots of womb massage but it can be done! When the uterus is way back, you may have trouble getting pregnant, and you may have really bad cramps and even fainting during your menses. If it's just low or to the side, we can easily fix it with a little womb massage and then you will feel much better and the cramping will be gone. You will feel so good with your uterus in position that you might say, "Oh, there I am bleeding and I didn't even feel it coming!"

Here in Belize, we have a mango tree called the black mango. When we feel a little bit bloated, tired or have a headache on the first day of menses, we take three mango leaves, boil them in a little water and sip the tea as hot as we can. Soon, we see some big clots stringing out. The mango breaks down the clots. Sometimes we use chamomile tea, again as hot as we can handle it. It's simple for us to eat fresh red hibiscus flower. We just go out and collect the beautiful flowers, which grow all over Belize, and eat as much as we can. We also use warm sea salt water to wash our abdomen and soak our feet. Remember, we don't bathe in cold water or drink cold things. All these things help relieve the symptoms, but until the position of the uterus is corrected, that problem is not going to totally stop.

Some people say it is normal if the uterus is tilted a little bit and that we have to learn to live with it. That's a lie, a total lie, because that is a big problem that we don't have to live with. Every woman deserves to have her uterus in a good position and to be happy and healthy. A uterus out of position will lead to problems down the road, whether it is cramps, irregular periods, fibroids, infertility or even cancer. It is best to get your womb all centered now and have a happy life!

When we get the uterine position corrected and drink our *tonicas de las mujeres,* (women's herbal tonics), we will be so healthy and fertile. Watch out! You could get yourself pregnant at any moment. This is where condoms come in. Condoms are the best form of contraceptive, because an IUD or anything inside your body can cause you problems. Anything inside you may cause irritation and inflammation, which leads to imbalance and disease. Condoms don't hurt at all and they work very well.

In our Maya ways, we are taught not to have sex during our menses because that gives away our power. Menses is a time of great power for women and we need to keep our power for ourselves. It's possible for a menstruating woman to pick up negative energy from a man when she has sex with him because she is so spiritually open during her moon time. Also, we believe that when menstrual blood mixes with semen that instead of conceiving a baby, a strange ball of flesh develops with hair and teeth and it looks like a little monster. In English it's called a molar pregnancy. That is something best to avoid.

For us, menses is a time of being with ourselves. It is a time of prayers and powers. We honor our sacred powerful energy. We do not give it away to some guy, even our husbands. It is a time for rest and dream visions, when the grandmothers cook for us and even bring us little gifts. Menses is usually not a time to enter into a big ceremony with a mixed group. A menstruating woman is very open spiritually. She can have her own ceremony with herself or other women that she knows and trusts.

If a woman comes to me with delayed menses and I am certain she is not pregnant, I check her stress level. Sometimes if she is too stressed, that's why her period is delayed. I give her ginger tea made with three big pieces of ginger or allspice tea made with nine little berries crushed up really well. I boil the herbs in a quart of water for five minutes and she drinks them warm. This helps bring on her menses.

If a woman comes along with amenorrhea or infertility, I give her la sobada and check to see if she is happy. Women need to live in a happy environment. A woman with amenorrhea may be troubled. She may be worried about being too fat or too skinny. She might be in a bad relationship. Perhaps earlier in her life, she may have been raped or experienced another trauma. Those memories linger in her body and she needs some healing. I might have her start with the let go ritual. Once she's comfortable, I offer her la sobada, *bajos* (a pelvic steam bath), herbs and prayers. For a good cycle, a woman needs to be happy and not all stressed.

For infertility, we do many things. We check the position of the uterus and the way the woman is living. Maybe she is not happy in her marriage. If someone is not happy, they may not get pregnant so easily. All those things I find out, so I can help her. The uterine position has a big role to play with fertility. If the womb is leaning on her colon, it is very hard to get pregnant. In that case, I give womb massage every week. I have a special trick of knocking the knees on a certain point nine times and I also give her herbs. She needs herbs that clean out the uterus and balance the hormones for three months.

I recommend she figures out what she needs to get happy, because if she is not happy her own body will tell her "no." She can take nervines and do her steam baths, too. She will need to take care of herself and do womb steams, womb massage, herbs, eat good food like fish oils, chaya, kalaloo, eggs, sea food, chicken and make any necessary changes in her life for three months to make a good place for that baby to come and live inside of her. After these treatments, she can try again to get pregnant.

To regulate our cycle we eat good foods like fish, veggies and greens. We make it a priority to live in a happy environment. We get la sobada, take our womb steams, drink our woman's herbs, eat fish and greens and live a happy life. That is how to regulate the menstrual cycle.

If a woman comes to my healing hut who has lost her periods altogether and is not pregnant or in menopause I ask her, "Have you been moving from place to place? Is your life very stressful? Might you be anemic?"

Of course, I give her la sobada.

If her period is simply late and she is for sure not pregnant, I give her a strong cup of chamomile, ginger or allspice tea and la sobada to get things going.

If a woman's feet are cold, it is almost certain that her womb is out of position and blocking a good flow of blood to her feet. I tell my female clients to avoid cold air or water, and no walking on cold floors or drinking cold things after a warm treatment. This causes cramps and an illness we call *pasmo. Pasmo* is caused by sudden changes of temperature.

If a woman comes to me with pasmo, her blood isn't flowing well and stagnant blood is the result. Maybe she went outside and got really hot while working in the sun in her garden and then she came inside, opened the refrigerator and had ice cold water. Maybe she went to a sweat lodge and then jumped in ice cold water. We Maya wrap up and keep warm and cool down slowly. When we get out of a warm bed, we don't put our warm cozy feet on a cold floor. Cold water is only okay to bathe in once you have let yourself cool down a bit. One of my brothers ate very hot potatoes and then jumped in the cold creek and got pasmo. We believe that if a woman wants to conceive, she should avoid cold things because cold eats semen right up. Women need to eat and drink only warm things. Cool is okay but it is best to avoid chunks of ice. A little bit of ice cream here and there is okay. Everyone is entitled to a little fun.

When a woman comes to me who is having chronic miscarriages, I can feel that her uterus is too low and cannot carry a baby. As the fetus grows, the too low uterus cannot hold it anymore. Everything, unfortunately, falls, including the growing baby. Before getting pregnant again, this woman gets a lot of womb massage and wears a faja to support her uterus. She eats really good food, drinks red hibiscus or red rose tea and other blood builders like nettles and red raspberry and doesn't wear high heels or lift heavy things.

I tell her, "Be happy and wait three to six months before you get pregnant again. Once you conceive again, take it easy. Don't work too hard or lift heavy things. Wait three months into your pregnancy and then start getting your pregnancy womb massage. You should avoid sweeping and vacuuming because those actions shake you up too much."

High heels are bad for women, especially when they are pregnant. In flat shoes, maybe you don't look so sexy, but the womb is happy. High heels throw the uterus out of position and cause back problems. Flat shoes are better and they can look just as cute. I think each woman is special in her own way.

If a woman is having difficult or painful ovulation, it means she is working too hard, is stressed out or unhappy. She needs to relax and have some allspice and ginger tea.

When fibroids are small, womb massage and herbs help immensely. Women with fibroids benefit from a fibroid formula made with herbs that both shrink and flush things out. Women with fibroids should also avoid cheese and oily things.

MENOPAUSE

When our child bearing years are coming to an end, we have choices, and we can finally do all the things we are inspired to do because we have a little more time. Most likely we have enough time because we don't have to raise the babies anymore. When our energy is free from child rearing, we can develop our own gifts and offer them to the community. As a senior citizen, you get a little compensation here in Belize. That gives you time to pick and choose what you really want to do in life. We get to a stage of having more wisdom, and we start talking about our life experiences with other women.

Still, we need to take care of ourselves. We need foods which are nourishing for our body such as amaranth greens which are rich in iron. We need to eat red chard, sea food, kelp, nuts, carrots and beets because beets enrich the blood. Here in Belize, we eat chaya and amaranth and wild spinach. We have beets and sea food and river fish because the Macal River is close to us and we have a market on Saturday. In menopause we take cod liver oil, but if you take it and you notice that you have a skin condition, stop for a while because cod liver oil creates heat. We only take just one capsule per day, especially in a hot climate. Eating whole fish is good, but to get a little bit of fish oil is even better.

You can eat free ranging chicken, red meat and lots of greens. You have to eat foods rich in calcium to prevent osteoporosis. In menopause, it is best to avoid coca cola, which will steal calcium from your bones and make them brittle. Cayenne pepper is good for your heart. It stimulates circulation and thins the blood. But too much cayenne, *chiles,* ginger and allspice are too hot for menopause. They will aggravate your heater and make hot flashes worse.

As you approach the end of your menstruating years, you might have fevers, achy joints, night sweats, anger, nausea, loss of appetite, feel picky about food, lose your desire for sex, have a dry vagina and have your hair fall. Your pattern will be similar to that of your mothers. All of this can be easier with a good diet, womb massage and herbs.

If menopausal women don't take care to nourish the blood with herbs and good food, they catch sickness like hypertension, diabetes and stroke. These diseases are also caused by lack of self care earlier in life and the stressful things that have happened to you in your life. If we had a lot of trauma in our lives, we have to try even harder to take care of ourselves. If we don't take special care, little by little we may encounter disease, not all of a sudden, but little by little, the stresses we have in our life affect our health.

In menopause we need to take nervines to calm our nerves. You could use man vine, skullcap, chamomile, valerian, blue vervain or passion flower. We also need to stick to our women's tonic herbs. We need to make sure we have a good flush before menses is totally finished. Every week, go and get your womb massage, la sobada and that will be the answer to your problems. Womb massage will help ease the symptoms you go through as your periods get fewer and farther between. Don't wait until things get worse.

Even if you had a hysterectomy, the womb massage will help you because it breaks up adhesions caused by surgery. Without massage after a hysterectomy, the woman may end up with heel spurs due to the lack of circulation through her pelvis and down her legs. Circulation is not good when the tissue is bunchy from scar tissue. She could get orthopedic shoes from the doctor, but that will treat the symptoms, not the cause. Only womb massage treats the cause.

At around the age of forty, we are usually entering the pre-menopausal stage. We Belizeans usually reach forty-five years old before we start changing. Pre-menopause is a huge change in your body, but don't think that at this stage you cannot get pregnant! Of course you can! With the last little bit of blood, you can get pregnant. So, you better put on two pairs of *chonies,* (underwear) especially if your guy is drinking a men's tonic!

Okay, when I was fully breastfeeding, that was my contraceptive, but still I was really afraid to get pregnant again. When I was thirty-nine, I had Zena and I promised myself not to have any more babies. I was worried because I could see in Zena's umbilical cord that I would have four more babies and I didn't like that idea so I made sure that was not going to happen! That is why I breast fed her fully for four whole years, because I didn't want any more babies. I took the task of breast feeding her as much as possible until she was four and a half years old. After that, I was premenopausal. I was achy, tired and sleepy.

I announced, "I am ready to bypass all these troubles!"

So I took my nervines and my woman's tonic for three years on and off. I ate my chocolate and after three years I didn't have a period for three months and then, for six months I couldn't see a period. Then it never came back. That was a relief.

I never had a painful period in all my life. My cycle was always regular, but at the end, yes, six months went by no period and then one year and then good bye! I said, "Thank you! My job is done!" I got way into my nervines which are herbs that help erase the troubles in your mind and calm you down. I took my grandmother's woman's tonics and took my steam baths, so that helped me a great deal. For three years, on and off, I took my herbs and it got to a point where I was very emotional. If a hen gave me a bad look, I was crying! Everything made me cry! Then came the hot flashes, diarrhea and achy joints. My mother went through it for three years. It is said that what happened to our mamas in menopause is what happens to us. I think it would have been much worse if I didn't have my Maya medicine and my prayers to help me through. At that time I got separated from my husband, which made all the stress double. I kept drinking lots of nervine herbs, ate my chocolate, did my steams, and got my massages. I also drank my woman's tonic and finally I got through it. When I had groups come to study with me, I asked the ladies to massage me every day. That is how I took special care of myself.

In menopause we could get rashes, dry vagina and age spots. Mayas think age spots are caused by blood toxicity so we take a shot of castor oil every year in menopause to flush everything out. If you have skin conditions, don't drink ginger or allspice tea because it creates heat in your body and will make the skin condition worse.

What we should do is take things that are cooling like lime, coconut and hibiscus flowers. Another thing that puts too much heat in your body is the kind of cloth you wear. If you have too much heat, do not wear nylon, just stick to cotton. All the little things like that need to be addressed. We also need to take bee pollen during menopause but bee pollen is a little bit dangerous, so I tell the ladies if you take bee pollen, mix it with honey because honey is the antidote. If you are okay after twenty minutes of taking a tiny bit of bee pollen mixed with honey, then you are okay and you can go ahead and take more. Not everybody can handle bee pollen; sometimes it causes inflammation and allergic reaction.

There are other things you could do like drinking wild yam and other herbs to balance hormones. Do not take the pill made out of horse urine because there are natural things you can do for hormone replacement that will be much better for you in the long run.

We have to try to have eight hours of sleep at night. Your energy will be drained out more easily if you have no sleep. Nervine herbs will come in handy in menopause because your worries come and go and nervines help you forget about all your troubles and get a good sleep. It doesn't mean the problems will change, but if it doesn't bother you anymore, then that is good enough.

Around menopause, sometimes we develop lumps in the boobs or feel bloated. We may feel bloated at the same day of the month that we used to have a period. Your body remembers what was happening, so when we feel bloated, we can drink chamomile tea and that will help. If we have a breast lump that we know is not malignant, we take a castor leaf with castor oil on it and put it on the lump. I had my little lump and I treated it with castor and after one day it disappeared. Many women have had a little lump. Most of the bumps are not malignant and can be treated with castor oil. Your health practitioner can tell you whether your lump is malignant or benign. If the lump is benign, we treat it at home with castor and a little ceremony. If it is malignant, it is time to go to the doctor. The simple small benign ones, like a cherry tomato, now we can deal with that! However, if the lump is malignant, or it gets painful, achy or red, go to the doctor and check it out.

The Mayas believe that when we have troubles, like breast lumps or other things that we want to be free from, we think about our father sun. He is so strong, he can take away all these problems. We brush away our problems at sunset. It is a little ceremony we do to release things we no longer want. The sun carries away all those things while he is setting. This works because we have faith.

Black copal wards off evil. We burn it and light a white candle to ward off ghosts and evil spirits. We ask the bad things to go away and the white candle invites the white light into our lives and into our homes.

I made a little ceremony when my menses ended to say, "Thank you for all my babies and the years of fertility." I said, "Thank you for helping me make it this far." I lighted my white candles and burned my white copal. I offered the corn porridge, white flowers and prayers to show my gratitude.

THE WISDOM YEARS

Even when a woman is finished with menopause, still she needs her womb massage, la sobada. La sobada is for all ages of women. It will always keep you healthy. It is something you have to have. There are some ladies who have many troubles, but if they get la sobada, they bypass the trouble and live happily. Hysterectomy could often be by passed by having womb massage regularly. Even after surgery, there are adhesions that need to be addressed with this massage.

This work needs to be passed on so women can be aware of what they can do for themselves. In America, many women have had gynecological problems going on for a long time but little by little they can be helped with massage. Better yet if they get womb massage from an early age, they can bypass all that trouble. Prevention is much better than trying to fix all the problems caused by a uterus that has been out of place for decades.

Gastrointestinal problems can arise as we age. They can also happen anytime in our life. In Maya medicine, we describe three types of gastrointestinal problems. One has bloody stools, one has mucous stools and one gives you diarrhea. Farting a lot is a whole additional problem. It is caused by wind in the belly accumulating over a long period of time. Headaches can accompany this gas problem as well. For farting we use fennel tea. Fried foods, cold foods and vinegary foods make gastrointestinal problems worse. We don't recommend eating cold or icy foods. When a woman is constipated, often her uterus is laying backwards in her pelvis, blocking a good flow. For gastrointestinal problems, we offer a special belly massage and particular tummy herbs along with calming herbs like blue vervain and warm drinks which warm and soften the stomach. Sometimes when receiving the massage and herbs you may fart a lot for a few days, but then it is all out and you will feel much better.

I have told you many things we can do for a woman's health through-out her whole life, but the most important thing is la sobada, our womb massage. I think every woman should know about it because it helps so much. Anyone who is ready to work at it and practice a lot can learn it.

Part of the learning is the knowing of when not to do it. No one should massage any uterus when it is bleeding. The bleeding after childbirth is different and a midwife can give a womb massage then, yes, but not during menses. After a big operation, no, you can't do any womb massage for a good long time. But after waiting until everything is healed, it will help a lot. Don't do belly massage after a huge meal, because of instead of doing something good, you could do something bad. You could tear the fascia and that won't help anybody.

We have to ask a lot of questions first before massaging a woman's body. Also you can never massage someone's uterus if there is an IUD or cancer. La sobada cannot be learned overnight. You have to learn from someone who really knows it and then you have to practice it over and over and find your own uterus first so you know how it feels, then you can practice on other women. You need a lot of hands on training. You don't have to be a Maya to do it. You just have to know what you are doing so you don't harm a woman's body. We need to respect women's bodies. So take care to learn well and always stay aware when you are working with people's bodies because they are entrusting themselves to you and you have to do your best.

Maya Plant Ally Number Five: Blue Vervain

Stachytarpheta cayennsis

Wild Blue Vervain on Caye Caulker, Belize (photo by Katherine Silva)

Blue Vervain is an important calming plant ally especially for women. It calms the nerves and relaxes the stomach. I drink blue vervain as a warm tea or in a tincture form in a little bit of warm water. Blue vervain's flowers are good brain food. I prepare an herbal bath, which includes blue vervain for new mamas after childbirth. For a swollen uterus, I give women a vaginal steam bath with basil and blue vervain. Its bitter taste makes it good for de-worming and helps dispel negative energy. Blue Vervain grows wild all over Belize with its beautiful blue flowers dancing in the breeze. It is a magical herb for women. A foot bath made with blue vervain is one of my favorites for easing away the troubles of the day. Ahhhhhhhhh!

Chapter Six: Our Sacred Waters

The Macal River near Miss Beatrice's Home (Photo by Katherine Silva)

"A long time ago, the Maya people who lived by the sacred waters would make a ceremony dedicated to the lord of the water. The lord of the water, who is the benevolent spirit that takes care of the water, would be pleased, and so the people always had enough sacred water."

The Maya respect all water, but there are some waters that are considered especially sacred. Any water that runs from a spring is considered by Maya people to be sacred water. Spring water is the best water to make your herbal baths with. If you live by a little creek that is running, you are very fortunate to have your healing medicine right there.

We show our respect for water in the way we treat natural watery places. If you see a little creek, you can check to see if there is a spring from the mountain, and if so, there will be little snails. If you pick snails, you have to be granted permission, just like with herbs, and you can only take a small amount. Just pick the big ones. All the small ones must be left there to grow. It's bad luck for anyone who just goes and picks all the snails. The ones that watch over the water will be angry. You have to leave something so the snails can continue to live and so there will be snails for next year for the people as well. So, it is best to leave the medium and small ones there for future generations and for yourself. We never pick all of something at one time. We always leave plenty of whatever it is to thrive there and flourish.

Also we do not bathe inside sacred waters, because we've seen what happens; it will eventually dry up. We can take water out of a spring and pour it over us for a blessing, but we do not get totally in the sacred water because that is disrespectful. That is not good. Because we find that by doing so, the water dries up little by little.

When I was much younger, we had a creek not far from our home, a little spring. It was a long time ago. We also had a tree of white copal and a big tree of allspice, and when it was raining, the little creek was flowing a lot, but in the dry season it was small and shallow. There were deep little pools where we had crayfish and many good-sized fishes, so when we would go to get the firewood, I would always take my little nylon bag. I like to go fishing. So when I would bring home the firewood, I would bring home some fish also, maybe ten or twenty, not so big, but good sized. My mother would make soup or we would clean them good, wash them, roast them and put a little bit of lime juice on top and add pepper and salt. Oh boy, what else could you want? It was a delicacy.

Then something bad happened to our sacred waters. A guy got so greedy. He bought that piece of land for his cattle and felled all the trees that were near the water, even the copal tree, even the allspice! My father said that foolish guy has done this and now that pool of water is gone for good. Only if it rains, the water will be there a little bit, but otherwise, it will dry up.

Also, his wife went and washed in the pool, and his kids bathed in it, and sure enough, that little pool full of water went dry.

My father said, "From way back, that little pool was always full of sacred water and never went dry. But as soon as the greedy man showed no respect, it went dry for good!" My father told us, "Water like that is sacred. Nobody should go and bathe in it. Nobody should go and wash in it."

A long time ago the Maya people who lived by the sacred waters would make a ceremony dedicated to the lord of the water. The lord of the water, who is the benevolent spirit who takes care of the water, would be pleased and so the people always had enough sacred water. But that guy got greedy and crazy, and felled down all the trees, even the copal tree. I tell you people get so greedy. He only wanted to see a clean place with no trees. What do you think is going to happen? Erosion–because all the trees have a purpose! The roots hold the earth and the leaves make shade for people and animals. Trees give our lungs good oxygen to breathe in. Big trees are homes for little birds to live in and have their nests, but he didn't think about any of those things, did he? He just wanted his place clean and that was it! He only thought of himself. He was very selfish, I would say. And now what is left? Our sacred waters are gone and so are the little snails and fishes which used to live there.

My father told me a story about sacred water. He told me that a little farther up from where we live, way, way back in the rain forest, a long time ago, was a small village where people were living. There was a little natural fountain where the water was always abundantly flowing. All year round, the people had clean water. It came from under the earth. It was sacred water. The people honored the water with their ceremonies and treated it with respect.

Then a greedy family took the land and they bathed in the fountain. My father told us that very soon that sacred water was gone. All of a sudden, it went totally dry.

He said, "So you see, when you don't take care of your sacred water properly, you are going to lose it!" Now some people have a farm there and they have no water.

In my father's *milpa*, there happened to be a little pool of water like a little vase made of stone. My father believed that the Mayas way back had made it. He never showed it to anybody. He said that little container is always full of water.

He told me, "Sometimes when you go there, it is full of dry leaves, but if you clear away the dry leaves, you can see the clear crystal water coming out underneath. "Who fills it?" my father would ask me.

"It is always full with water, so who fills it? That fountain is close to a big mountain and nobody lives there. Kind spirits are taking care of that water. It has to be, because nobody lives close by. So, that is why that water is sacred. It is for the birds to drink, so the animals can survive and even for human beings, because I drank from that water and I never got sick. But who fills it? I have no idea!"

He also said that when we find a big hole close to a pool of sacred water, the hole leads to an underground road which stretches all the way to the great Maya city of Tikal. That is the mysterious Maya connection.

So you see, these sacred waters will survive only if people take care of them. If no one takes care of the water, if people disrespect it and neglect to offer their ceremonies, it will dry up for good.

Maya Plant Ally Number Six: Rosemary

Rosmarinus Officinalis

Mama and baby enjoy an herbal bath (Photo by Monica Juchum)

Rosemary must be included in any talk about plant allies. She is cleansing, sacred and protecting and an important ingredient in many of our Maya herbal baths and teas. Rosemary is a good one to put in the postpartum womb steam bath and to wash the lactating breasts to help bring the milk down. We use rosemary to bathe with, or to burn as incense after visiting a cemetery. Rosemary is burned right before the birth of a child and in the baby's room where he is sleeping with his mother, as a way of blessing and protecting the child.

Miss Beatrice processing Cacao (photo by Terra Rafael)

"My mother used to harvest and make her own chocolate from our own chocolate tree. When she mixed her cinnamon, water and chocolate together with some wild honey or cane sugar (grown and made by my father and grandfather), it was so good that the more we drank the more we wanted!"

SALSA

In our house, we have three kinds of salsa. One you boil, one you roast and one you make fresh. Every day we can have a different salsa.

Roasted Tomato Salsa

You will need:

1 large tomato

½ sweet (bell) pepper

1 clove of garlic

1 bunch cilantro or *culantro* (a wild herb that grows in Belize)

¼ large onion or a handful of chives

Salt to taste

Optional: ½–1 *jabanero,* bird or j*abanero* pepper

Take your tomato and roast it on the comal for twenty minutes or until the peel is slightly darkened but not burnt. Let the tomato cool and slide the skin off. Chop the roasted tomato finely or throw it in the blender. To the blender, add your garlic, cilantro or culantro. Culantro is a wild herb that grows in the Belizean jungle. It is similar to cilantro but a little more spicy. Add your onion or chives and if you like it spicy add your hot chile pepper. If you want it medium, just add half of a pepper; but if you like it really spicy, put in a whole pepper. If you don't have a blender, you will have to chop everything finely and mix it together. The last thing is to add salt to taste. This salsa is delicious with tamales, roasted meat and soup. It has a special taste because it is roasted. We just had some yesterday. It is very, very good!

Boiled Salsa

You will need:

4 medium tomatoes

2 cloves of garlic

A bit of olive oil

½ to one chile pepper

Wash your tomatoes properly and boil them for ten minutes. Skin them and set them to cool. Do not throw away the water in which they were boiled. When cool, put your tomatoes in the blender with a splash of the cooking water, your garlic, your oil and your hot chile pepper. Blend up all the ingredients and then salt to taste. It is very important to put the salt in last; otherwise it gets too salty for some people. If you don't have a blender you can chop everything finely, then smash it all together with a mortar and pestle. If the salsa tastes too watery you can put a little bit of olive oil in a pot and slowly cook your salsa until it thickens This salsa is scrumptious with bols, on top of meat or fish, inside tamales or eaten simply with homemade tortillas.

Miss Beatrice's own Fresh salsa

You will need:

4 tomatoes

¼ onion

1 sweet pepper

1 bunch cilantro, *culantro* (a wild Belizean herb) or basil

1 pinch black pepper

Juice of 1 lime

1 pinch salt

Cut up your tomatoes and onion into small squares. Cut your sweet pepper a little smaller than the onions. Chop the basil, cilantro or culantro very finely and mix it all together, adding your lime juice, salt and pepper. This salsa is good with everything!

Caldo Cum (Pumpkin Soup)

You will need:

1 ripe pumpkin

A handful of fresh pumpkin or squash flowers

2 ½ cups of water

1 bunch cilantro

1 green pepper

Pinch of cayenne

Pinch of black pepper

1 pound of *masa harina*

1 tablespoon olive or coconut oil

If your pumpkin is older, you will first need to peel off the outer shell and remove the seeds. If it is a young pumpkin, you can use it whole. Either way, cut the pumpkin into little squares while you boil your water. Add your chopped pumpkin to the boiling water and cook for ten or fifteen minutes or until soft. When the pumpkin is soft, add in your flowers, cilantro and green pepper diced very small. Now add your salt and pepper to taste. The last part is to add your dumplings.

In a separate bowl, pour in your masa and add to it a pinch of salt, and one tablespoon oil. Little by little, add water to make a dough that is not too soft and not too hard. Form little tortillas with this dough with your hands and gently drop them into the pumpkin soup. When the little tortilla dumplings are cooked, the soup is ready. This is a very healthy food. You don't even have to make tortillas because everything is already in there. The soup is lean and nutritious not to mention delicious!

Amaranth greens (Kalaloo in creole) (Queltosh in Yucateca Maya)

You will need:

1 pound fresh amaranth greens

1 small onion finely chopped

A splash of olive oil

2-5 cloves of garlic finely chopped

Water for boiling

Harvest your amaranth leaves with respect and gratitude. This is a very special plant. Boil your amaranth leaves for a little while to soften them, then chop them up saving the water in which they were boiled. Cook your onion in a large cast iron pan over a slow fire in a little bit of olive oil for five to ten minutes. Now add in your chopped amaranth greens, your garlic, your pepper and your tomatoes or tomato paste. Mix it all together while you cook it down, adding the water as needed. When everything is soft, well blended and cooked through, it is done.

Amaranth greens, the fresh green leaves, are rich in iron and good for your nervous system if eaten three times a day. It is said to have properties of insulin so we call it "insulin provider," but you have to eat it on a regular basis to get the benefit. Amaranth greens are very helpful for someone with anemia. When you taste a little bit of it, every day you want more and more and more! It triggers something inside of you to make you want to eat more of it. Kalaloo is served with tamales and fried plantain or all on its own.

Amaranth is a very special plant. When you chew the seeds, it tastes a little like blood and that tells you something. It tells you that amaranth builds your blood and gives you energy. A long time ago, amaranth was banned in Mexico because the Mayas used the amaranth flowers on their altar. The officials said that nobody should use it because it's against the law. When the officials burned all the amaranth crops and forbid anyone to grow or use it, The Maya secretly collected the seeds, put them in cow horns and buried them deep in the ground. Thanks to them, up to this day, we still have amaranth. The officials back then were scared of amaranth's magic and power. It is true that amaranth is magical and powerful, but only because it is very nutritious. Its magic is that it nourishes your body.

Real tamales made with plantain leaf

You will need:

3 pounds masa harina

2 onions finely chopped

1 head of garlic finely chopped

1 cup currants (optional)

1 bunch cilantro or chives finely chopped

2-3 pounds beef, pork or one whole chicken

3 cups chopped veggies of your choice

1 pound lard or 2 cups oil

½ gallon water

1 Pinch salt

3 cans tomato salsa or your own boiled salsa with added onion

Many large plantain leaves

(The making of tamales is a family event reserved for special occasions such as birthdays, confirmations and ceremonies. The entire process of gathering all the ingredients and preparing them with love, takes many helping hands two to three days).

First you have to gather your plantain leaves, run them over a hot flame and wash them by passing a clean wet cloth over each of them to wash the ashes away.

Mix your water and masa harina so it is not too thin and not too thick.

Put your oil or lard in a big pot over your fire heart and add masa, water and salt to the pot. Stir and stir it very thoroughly until it is smooth and bubbling on the surface

Set masa aside to cool while you cook up your meat, veggies, onions, garlic and your cilantro or chives. The meat should be cooked with plenty of water just like we make "stew meat," to keep it juicy.

Lay out your plantain leaves on a piece of tin foil, double them to hold the ingredients well and fill the center of the plantain leaf with a big spoonful of your cooked masa. Then cover with the fillings of your choice. You can make different combinations with meat, veggies, onions, cilantro, garlic, chives, salsa and currants. Then fold it up and steam in a steamer for twenty minutes

At home, we make a steamer by making a grid on the bottom of a big pot with plantain stems and putting the tamales on top of the grid with three cups of water underneath the grid, nearly touching the tamales from below.

These tamales are so good that you will want to eat so many of them!

Bols

You will need:

1 medium sized carrot

2 pounds masa harina

¼ cup olive oil

1 ½ quarts water approximately

1 sweet pepper

One bunch of chaya, amaranth greens, wild spinach or collard greens chopped fine and cooked

½ pound cheese (optional)

1 cup pumpkin seeds ground

Pinch cayenne (optional)

Soak and cook up beans with lots of garlic. Garlic makes the beans tasty and easy to digest. If you have grown your own fresh beans, that is even better. Peel them and set them aside. Then you get your masa harina or your own masa that you have grown, cooked with lime and ground. However you get your masa, cook it in some water until its not too thin and not too thick. Do not strain it. When the masa is cooked and smooth, mix the masa properly with the beans, greens and oil and a pinch of salt.

Grind your pumpkin seeds and cayenne pepper and set to the side. If you don't like hot pepper just bypass it. Grate some carrot and chop up some sweet pepper and put it to the side.

Now you are ready to prepare some plantain leaves. Pick them with reverence, pass them over a hot flame, rip into square pieces, about ten by ten, and wipe them down with a clean cloth. If you don't have plantain leaves you can use corn husks, but plantain leaves have a better taste than corn husks and are much easier to work with. You have to soak your corn husks to make them soft before you use them.

Now mish mash the masa and bean mixture and make little tortillas with it. Put your little tortillas on top of your plantain leaves or, if using corn husks, just stuff your mixture in there. Now add in some cheese if you want and then sprinkle your pumpkin seed and chile mixture on top of the little tortillas. Add some sweet peppers and grated carrot, if you wish, and roll it up! Corn husks are much more work then plantain leaves. Put your bols in a steamer for forty-five minutes. If you used corn husks, let them be standing up so no water gets in while steaming.

When it is done and it is opened before you, oh boy! It could be accompanied with meat if you want or by itself. This dish is very common in Guatemala.

Fish Soup (Caldo de pescado)

To serve four people, you will need:

2 medium sized fresh fish

Handful okra chopped

½ of an onion finely chopped

3-4 cloves of garlic

Pinch ricato mixed with water

½ glass of water mixed with 2-3 spoonfuls flour

2 tomatoes

1 small sweet pepper

1 young pumpkin or 1 yam or a carrot or other root vegetable

½ gallon water

Pinch cumin

Pinch black pepper

Pinch salt

Bunch cilantro or culantro

Clean your fish very well and wash it in a little piece of lime. Have your big pot ready on the fire. Fry your little piece of chopped onion in olive oil in the pot; then add your water. You then put potatoes or pumpkin or other root veggies in there in little pieces and when that all is cooked you add a little bit of flour so it doesn't taste too watery. The flour makes it thick. Then you need to add garlic, cumin and black pepper and let it boil. When the vegetables are cooked you can put your fish in there whole. If you want coloring, you can put a little bit of coloring like ricato, but just a pinch. At the very end, dump in your okra, so that the soup doesn't get too slimy. Then you are ready to have your cilantro or your culantro chopped in pieces to spread on top. Let it simmer for ten more minutes and then you have a nice soup you can eat with corn tortillas, with rice or with yams. It's very good! It's so good you could eat nothing else!

Pirish Pacque

You will need:

2 large tomatoes

1-2 local eggs

½ of an onion

2-3 cloves garlic

1 Jalapeño or sweet pepper (optional)

Splash of oil or water and Salt to taste

How much to make depends on how large your family is. Let's make this one for only two people. You take your two good sized tomatoes and chop them in little pieces. Then you put your pan on the fire with a little oil or water, whatever your taste bud tells you to put in there, and cook your tomatoes that way. If you want, if you are someone who likes spice, you could chop your jalapeño pepper in very small pieces and cook them with the tomatoes. If you don't like spice you could bypass that, and you could put half of a sweet pepper in instead. In another pan, fry your onion. Then dump all that tomato in with the onion, and when it has really cooked, add in a little pinch of salt. Break two or three eggs in a container first to make sure they are good and all clear. If the eggs look good, put the eggs on top of the onion and tomato mixture and cook really slowly on medium-low fire until the eggs are cooked. On top of the eggs you sprinkle a little bit of black pepper and a little pinch of salt to taste. The whole time it slowly cooks, kinda stir a little bit here and there and then leave it to cook. Then you mix all the tomatoes and onions with the eggs again and and that's done. It's so good! Yes! Pirish Pacque is eaten with corn tortilla or flour tortilla. It is a delicacy.

Sour soup (escabeche, onion soup)

You will need:

1-2 cloves garlic

1 onion

1 whole chicken

6 sour oranges or limes or lemons

Pinch black pepper

Pinch cumin

Pinch allspice

Pinch oregano

Get a chicken, a local one, undress it and wash it and boil it until it is soft and save the water for your broth. Take the chicken out and roast it. Put the juice of six sour oranges or limes or lemons into the same water you boiled your chicken in. Cut the roasted chicken in little pieces and add it into the broth with black pepper, cumin, garlic and allspice

Then you need to add in onion chopped in little round pieces. Lastly, you get a little bit of oregano and put it on top and simmer covered for ten minutes. Sour soup goes with rice or corn tortillas. If you leave a little bit for the next day, oooh you want more of that! The older it gets, the better it tastes! It's sooo good!

Home Made Corn tortillas (tortillas de maiz)

Corn tortillas we eat every day. Corn tortillas start with masa. To make real masa you cook your own corn with white lime in a covered pot. Then when it is cooked, let it sit a little bit and grind it up and there is your masa flour. If it cooks too long it becomes hominy. That is not good for making tortillas. Or you can buy some masa harina if you like. However you get your masa flour, wet it with warm water and make it thick. Make it into little balls and then press it round and flat with your hands. Cook them for a few minutes on the comal, turning them every 30 seconds until lightly browned.

Flour tortillas (tortillas de harina)

You will need:

1 pound white flour

1 teaspoon baking powder

Pinch of salt

Small amount of vegetable lard

Less than ½ cup of water

Mix white flour and baking soda in a bowl and add your pinch of salt. Add water little by little until dough forms-not too dry and not too sticky. Pull and stretch the dough, adding veggie lard little by little to soften the dough. Let dough sit ten to fifteen minutes before making tortillas. Prepare a smooth place to press your tortillas. Put the comal on the stove to get hot. Put a small amount of lard on your smooth working area. Roll dough into little balls and press them with your hands and then place on smooth working surface to further press dough until it is flat and circular. Cook your tortillas on the hot comal, turning them every fifteen to twenty seconds until fluffy and lightly browned.

Miss Beatrice's daughter, Marlyn, making tortillas (photo by Katherine Silva)

How to prepare chocolate for the Maya Gods:

The chocolate is collected from the chocolate tree with prayers and then taken from the pod. It is put to dry in the sun for at least three days. Then it is ready to roast. It is roasted and ground on a stone or grinding machine. Take three cups of water and two tablespoons chocolate, dissolved properly in warm water. Then add honey and or milk to taste.

A Word about Chocolate

Chocolate is very special to the Mayas. Mayas consider chocolate to be food for the Gods. Way back, chocolate was always served as a drink reserved only for special occasions. Back then, chocolate was also used for money because it was so special. Chocolate was only for noblemen, kings and rulers; not just anybody was entitled to have it. If you got some chocolate, you were very lucky. Chocolate is a medicine, too. It really helped me all the way through menopause. It is nutritious and it is an aphrodisiac.

My mother used to make her own chocolate. My father had a chocolate tree and a coffee tree. The chocolate tree died, but the coffee tree is still living. When the chocolate tree was living, and it was time to harvest chocolate, my mother would pick the chocolate pods with prayers and smash them in half. When chocolate is ripe, the outer shell has a juicy part that is both sweet and sour. That was the job of us children, to suck on the pods. When the juicy part was all sucked out, my mother would sun the chocolate pods. When the chocolate was all sunned (dried), she would roast it and peel off the outer layer.

Then she ground it up on a stone. After a while, she got a grinding machine, but way back, it was done on a stone; everything was done on the grinding stone. When it was ground, and the grinding stone takes a good while, it was really smooth. When the chocolate is done and really smooth, you can touch it and see the oil seeping out.

For us, chocolate is used only for ceremony or once in a blue moon, for a special occasion, like an engagement or a wedding. So when my mother harvested, roasted and ground chocolate, she would make it into little round balls and keep it in a container and save it for when she really needed it. When the special day would come when she was going to make some chocolate, she would get a container with warm water and take two little round balls of chocolate out of the jar and put them in the warm water. She would mix it with her hand or with a big piece of stick with five little twigs sticking off of it that mixes and mixes and mixes. Once it was really mixed, she would put it in a big pot of really hot water with cinnamon sticks inside. When she would put it to cool, we could see all the yellow fat rising to the top. She collected all that yellow fat to make chocolate fat salve (coco butter). That salve is really useful. It removes scars and pimples. She used the chocolate fat to cook food with too. When you cook with coco butter, it is soooooo good! You think "Oh! I'm eating chocolate!" But, it's really only the fat of the chocolate you are eating. It's really good! We make a recipe called Mole with the chocolate fat. When my mother mixed her cinnamon with water, chocolate, and some honey or cane sugar (grown and made by my father and grandfather), it was so good that the more we drank, the more we wanted. The chocolate they sell in the stores nowadays is nothing compared to the original chocolate like my mother made. The real kind, looks paler than the store bought kind but you can taste that it is much more chocolatey. The store chocolate may be brown, but it doesn't have the same rich taste.

Maya Plant Ally Number Seven: Rue

Ruta Graveolins

Miss Beatrice harvesting rue (photo by Janet Caspers)

Oh, I love this plant! Rue is like an angel showering light on everyone. Rue is used as a powerful protection from evil eye. Just put a bit of fresh rue under your tongue and no one can give you evil eye. We use rue for spiritual baths and spiritual healings and house blessings. Rue is specifically helpful in a bath in cases of grief. Rue clears away negative energy and is a very powerful herbal ally! Rue is an excellent plant to put in a protective amulet and to put under your pillow in a cross form to protect yourself from envy. You can also put a little sprig of rue in your wallet and more money will come to you.

Chapter Eight: Memoirs

Miss Beatrice (Photo by Monica Jochum and Janet Caspers)

"I was really surprised because she knew nothing of my history and had never met me before, yet had immediately understood so much about me." Nina Palmieri

"During my session, Miss Beatrice picked up my wrist and began whispering prayers into my pulses. In an instant, I was transported to a deeply comfortable place, as if I was being rocked in the arms of the mother goddess." Monica Juchum

Rocked in the Arms of the Mother Goddess

By Monica Juchum

The first time I met Miss Beatrice was when she visited Boulder one summer time. I think it was 2004. Ann Drucker had told me that Miss Beatrice was coming into town and that I absolutely needed to book a session with her because she was a fabulous traditional Maya healer.

Without knowing anything else about Miss Beatrice except what Ann had told me, I booked the session with her. During my session, Miss Beatrice picked up my wrist and began whispering prayers into my pulses. In an instant, I was transported to a deeply comfortable place, as if I was being rocked in the arms of the mother goddess. Wowza! That had never happened to me before! The experience was deeply significant for me. Miss Beatrice then very matter of factly told me some Maya remedies specifically for me. I was instructed to do some herbal baths and take some herbal tonics. Although I was deeply moved, I had no idea what she was talking about!

Soon after my first meeting with Miss Beatrice, I found a flyer that was an invitation to an evening get together to learn more about Maya cures and explore the possibility of joining a group that would travel to Belize to study with Miss Beatrice. After this evening, in which someone helped me palpate my own uterus, I decided to go on the trip to learn from Miss Beatrice. Two days before I was scheduled to fly to Belize, the leader of the group had a lot to say about riots and violence against women and backed out of the trip. My friend, Janet, was already there vacationing on Ambergris Caye and agreed to meet me there, so I went anyway. When we arrived at Miss Beatrice's house without the group leader and no money, the look on Miss Beatrice's face was priceless! She had no idea who we were! It was hilarious! Although she was not laughing at the time! The three of us proceeded to have an incredible time together. She's been in my heart ever since!

ETERNALLY GRATEFUL

By Nina Palmieri

It was suggested to me by a friend that I receive a session from Miss Beatrice because I had a tipped uterus and she thought Miss Beatrice could help me. I also had a history of major menstrual difficulties including severe cramps accompanied by vomiting and hospitalization, if not controlled through allopathic medication. I had chronic bladder infections and asthma as well.

When I arrived and began my session with Miss Beatrice, she whispered her prayers while holding plants on my wrists, forehead and heart. After that, she asked me a bunch of questions. She asked me if I had been in a car accident and if I had chronic bladder infections. She also explained that she sensed death in my energy field and also felt that I was holding on to fear relating to death. She said that she sensed I had asthma which was related to fear from a traumatic death-related incident. I was really surprised because she knew nothing of my history and had never met me before, yet had immediately understood so much about me.

Everything she had said was true. Yes, I had chronic bladder infections and yes, I was in a severe jeep accident as a young woman where I was ejected and hit a tree and the girl driving, my friend, had been killed. Soon after that accident, all of my health complications arose. Miss Beatrice suggested that my uterus was tipped and that it had adhered to my bladder. She said that once it was fixed I would have less pain and my flow would just flow smoothly. The flow she was referring to was my menstrual flow and also my flow of energy throughout my body. She said this would take three sessions.

As Miss Beatrice massaged my abdomen it was very painful to experience my bladder detaching form my uterus. It hurt so good because it felt so right. At the end of the session, Miss Beatrice said that the three sessions would create such a profound healing that would make such a big difference that I would remember her.

The next day I got my moon time so I wasn't able to continue treatment with Miss Beatrice, since she was only here a short time and she said she couldn't do the massage during menstruation. She referred me to Katherine Silva who was hosting Miss Beatrice's visit to Arizona. Katherine also said prayers with plants placed on my body while whispering prayers as Miss Beatrice had done and continued the deep massage on my abdomen. On my second session with Katherine, which made it my third uterine massage, I had an amazing experience. I felt a huge shift in my body and a major influx of energy. Within twenty-four hours I spontaneously developed burns on my wrists, my heart and my forehead. (Katherine later pointed out that those were the points where the herbs were placed while the prayers were said during the three sessions). Along with the burns was an enormous influx of light. At the time, I felt like the burns were a mark of God because up until that point, I had never felt so close to God as I had then. After the three sessions, I had a recurrence of my bladder, yeast and bacterial infections as well as a recurrence of asthma accompanied by a fever and severe nausea, which I experienced as a cleansing. This was an intense healing crisis, which eventually lead to a vast improvement in my health.

Although all of the physical discomforts were present, I couldn't deny my new closeness to God and that outweighed the discomfort.

From my first session with Miss Beatrice and the follow ups I had with Katherine, I had many memories and emotions come up relating to the jeep accident that were being cleared out. Also, my posture changed and my beloved said I looked more womanly. Most excitingly, I was able to majorly cut down on the amount of ibuprofen intake. I used to have to take at least twelve pills in a twenty-four hour period to prevent vomiting from the pain of my menstrual cramping and now I take two to three per monthly cycle. I feel that I experienced a miracle.

I was in the jeep accident when I was fifteen and had major complications since then. I tried multiple modalities of healing and at times was convinced it was all psychological because no one could tell me exactly what was wrong or provide a cure. To my delight, this treatment of womb massage and shamanic healing from Miss Beatrice and her student showed me that I really had been experiencing physical pain that was caused by my mal positioned uterus and which was rectified by this specific treatment. The physical, spiritual and emotional effects of my trauma were also addressed with the shamanic healing I was given. I actually do feel life flowing through me now without energetic or physical constriction in that area. I am eternally grateful and I do remember Miss Beatrice and send her all of my love and gratitude!

TOO NAIVE

By Janet Caspers

I never imagined or believed that I could be possessed by an entity. How foolish I was. Ann Drucker was leading a group to Belize for continued shamanic studies with Miss Beatrice. Prior to the trip, Ann wanted each of us to have experience "journeying." Since my prior experience was minimal, I thought I'd attend a little class offered through a bookstore in Denver. I didn't know the teacher or any of the participants. Each member practiced individual "journeys" while the teacher led the exercise. In my "journey," my power animal flew me to Belize where we circled Miss Beatrice's healing hut. I looked down and saw the hut with wild yam growing at the doorway of the hut. At that time, I didn't really know the significance of the vision I was shown. That information would come later when I was in Belize. Less than a month later, our departure date arrived. A few days before I was to leave, I started feeling ill. I took a lot of herbal remedies to keep the sickness at bay, but the evening before I was to leave, my conditioned worsened. I came down with a fever of 102 degrees and I ached all over. I struggled through the night but decided that nothing was going to keep me from seeing my beloved Miss Beatrice again. Early the next morning, I bundled my shivering body in fleece, and made my way in delirium to the airport. I remained in my hat, coat, and gloves until I got to Belize. The warm, humid weather in Belize helped rid me of my fever, although I couldn't sleep at night and just didn't feel like myself. Prior to going to Miss Beatrice, I did take an herbal bath and attend a *Primicia* led by another teacher. These treatments all seemed to help, although I still felt like my energy had been drained out of me. When the day arrived that we were to go and study with Miss Beatrice, we all climbed in for the half hour ride over bumpy, dusty Belizean roads to the little town of Santa Familia, where Miss Beatrice has her home and healing hut. We gathered in a circle in her healing hut for our class with her while she went over some Mayan shamanic practices. The time for the demonstration arrived. We were told to gather some herbs to use in the healing. Miss Beatrice had the fertile blue eggs. She would perform a spiritual healing with these tools. She picked out an herb. It was the wild yam plant from my vision. We were standing outside her healing hut and she was holding the Wild Yam plant. I told her about the vision that I had prior to coming the Belize.

She said, "Then we should do the healing on you."

We went further from her house to a more isolated place on her property in front of one of her guest huts. With the group gathered around, I sat in a chair and Miss Beatrice picked up my arm and placed the wild yam and blue egg on my wrist pulse and she began her prayers. The moment she started to pray, I started to shake. A blood curdling scream came out of my mouth. This horrific noise did not feel like it was coming from me, but rather from something surrounding me. It spit, it shook and it kicked. It made horrible growling gurgling sounds as Miss Beatrice calmly said her prayers into my pulses and in a cross direction on my body. It felt like this horrible storm was being released from me. Luckily, it was not deep inside of me. It felt like it was more superficial...it hadn't totally invaded me, but it was trying to keep hold. Miss Beatrice was both calm and commanding with her prayers. This entity was no match for her!

Finally, the shaking, hissing, and screaming stopped. The storm was over and I was totally wiped out. Miss Beatrice told the other students to take me to the hut to rest. A nurse in the group came and attended to me.

After she completed the healing on me, Miss Beatrice lovingly scolded me saying, "I told you Janet, you are *too* naive. You need to be careful who you go into a circle with. You got this from one person you were in a circle with."

She then cracked the egg she used to heal me with into a clear bowl of salt water. She would let this sit for a while, and then she would get more information from the floating egg in water. I lay there trying to think of who I had been in a circle with. The only group I could think of was Ann's group. Since Ann was very selective in choosing her groups, I knew none of these women had passed on the entity to me. Then I recalled the class I took in the bookstore in Denver. But, as I was running the scene through my mind, I couldn't think of who that "one" person was, because in that situation, there were actually two people. Late in the class these two very strange people came in to join the group. I did not like their energies at all. And yes, there was a point where we all held hands in a circle and said a prayer. So I *did* go in a circle with some people with very weird energy. Then we all "journeyed" in the same small room together. When one "journeys," one can be open and vulnerable. Those darkened energies must have taken advantage of me at that time to inhabit my energetic space. But to my good fortune, at the same time I was being invaded by those entities, my guiding spirits were also giving me information about the cure. The cure involved Miss Beatrice and wild yam. After I rested, Miss Beatrice came back into the hut to "read" the eggs. She told me that it wasn't one person, but two people that had done this to me. That confirmed my memory of that day. She was very firm when she told me that I should never "journey" in that fashion again. She told me that she did not "journey" in that way. She used her dreams for her answers from the spirits. When she has a question, she puts it to the spirits in her prayers, and then waits to get her answer in her dreams. She told me that I was *too* open to "journey" in a group like that. She warned me that dark energies *are* out there and they did like to attach themselves to healers. They can be so powerful that they tried using their dark energies to make me sick and keep me from going to Belize. But I remember my determination. Although I was delirious with fever, I felt that I could not be stopped from going. My willpower kept me focused on my goal-going to study with and be healed by my beloved teacher, Miss Beatrice. Since then, I have taken her strong advice to heart. I have never "journeyed" like that again.

SHE IS A TRUE HEALER

By Antonio Silva

I met Miss Beatrice for the first time in Taos, New Mexico, in June of 2003. She had come to do healings, ceremonies and workshops based out of our house. She did some healing work on me and insisted that I come to Belize for more. It was on my second trip to Belize that Miss Beatrice announced to me that there was a reason that I had to come to her house for more healing.

She said to me, "I could not tell you this before because I didn't have all my tools and my healing hut, but you have a serious curse on you!" She said that the curse kept me from succeeding at most endeavors and was currently destroying my marriage and relationship with my wife. With an incredibly intense look in her eye she said, "This person who made this curse wanted you to live like a dog!"

I chuckled and said, "Oh, that would be my ex-wife!"

Miss Beatrice then proceeded to tell me that the healings would take three days. I was staying at her home in Santa Familia and every day for the next three days, near sunset, she worked on me. After every treatment, she appeared very tired and exhausted. I started to wonder if she would succeed. At the end of the three days, she announced that she was finished, but mentioned that it had been very difficult.

Prior to having the curse lifted, when I would go hunting something would always go wrong or I couldn't locate the game. It was very unlike me to not have a sense of where to find the animals. Since the curse was lifted, many areas of my life improved, especially the hunting. Eight months after Miss Beatrice worked on me, I was invited to go hunting on a private ranch in New Mexico. I easily located a herd of elk and brought home a young bull. Two years later, I hunted again near my home and near Sedona, Arizona. I do not consider myself a trophy hunter and prefer to take young bulls, as they are more tender. However, on this occasion I was presented with a huge six by seven trophy bull and I took him home. I have his beautiful skull and antlers mounted in my living room. Whenever I look at him, I remember that my ability to hunt successfully was restored by Miss Beatrice's healing abilities.

During the time that I carried that curse, I associated and did ceremony with several Lakota medicine men and a couple of *Curanderas*. None of the these people had ever mentioned that there might be something wrong, although I was having frequent nightmares and many other things in my life were not going well. I have associated with many so called healers, shamans and medicine men and Miss Beatrice is, without a doubt, one of the most prolific healers that I have ever met. She is also rare among the ones who call themselves healers, as she does not refer to herself as a *curandera*. She has always maintained that she is a "*sobadera*"(a massage therapist). She has always been gracious and extremely humble. She is a true healer.

WITH ALL MY HEART

By Terra Rafael

Miss Beatrice has been one of my most powerful teachers. She gave me the most powerful tools for healing—prayer and laughter. Miss Beatrice taught me to pray again, as I had as a child, "with all my heart, with all my heart." My blessings for those I have cared for became more focused through prayer. She taught me to pray to Archangel Raphael, the patron saint of healers, happy marriages, and casting out of demons. These prayers helped me find my wonderful husband.

I noticed when I was assisting her that Miss Beatrice often would joke and get her clients to laugh. She encouraged laughter. I have followed her advice in working with women in my practice. The laughter lightens our burdens and allows us to be grateful for all that we have.

Miss Beatrice also helped me find my own name. Becoming close to Archangel Raphael convinced me to change my name, for once and for all, to Terra Rafael. Raphael means "God has healed" and God has surely healed me, with the help of Miss Beatrice and many others.

I will always revere Miss Beatrice as one of my teachers. Her picture is always on my healing altar, so that I can call upon her in working with women. May she always receive the blessings she has given us, multiplied by millions.

PANDORA'S BOX

By Sandra Lory, herbalist, Barre, VT, USA, July, 2010

I met Miss Beatrice Waight in Vermont ten years ago, at a workshop on Wheelock Mountain. I was the cook and got to sit in on some of her classes. And so I began my studies of herbalism and shamanism. We became quick friends and had a lot of great laughs. Later, visiting her healing hut in Belize inspired me to follow the path of being a folk herbalist, as my grandmother was. This is a choice not seen as a legitimate career path in over-regulated, computer technology and consumer driven, unsustainably disconnected from nature, polluted corporate America. Miss Beatrice helped me with my health issues and safely opened and released disturbances in what she calls "Pandora's Box." In that first weekend with Miss B, she touched me profoundly and stretched what I thought was possible in many realms of life and health. She rocked my world, and continues to do so.

Miss Beatrice's voice speaks up often in my mind, so clearly it's like she is in the room with me. She may be at home in Santa Familia and I can still hear her contagious laughter. Her feisty, good humored and reverent teaching style transmits her rich, lifelong experience into practical information that is so needed today. She reminds me to love my uterus and take care of myself in order to take care of others; to keep my feet and kidneys warm; to do herbal baths, steams and soaks; to hula hoop and dance and jump on the trampoline to keep my uterus in place; to drink beverages without ice; to rest the week before and the week of my moon cycle; to care for the ligaments that cradle my uterus by not lifting heavy things; to let go of stress; to perform limpias to clear a body or a space of unwanted spirits and energies; to eat bitters for good digestion and healthy blood; to love my parents and take care of my family and community; to trust the plants as the wisest healers; to listen to the messages in my dreams; to burn copal incense for protection; to speak prayers every day; to grow rue, basil and marigolds in my garden; to thank the angels, loving spirits and ancestors; to honor the seasons and elements of nature; to eat simple foods, especially dark leafy greens; and to address the spiritual imbalance of an illness in order to heal it from its root cause.

I cannot thank Miss B enough for the gifts she shares so freely with us all, and pray she is forever rewarded with abundance, love and support as a give back from her students and patients in Belize and worldwide.

THE CROW OF THE ROOSTER

By Fiona House

It was November, 2009, and I was heading to Mexico for a conference. I figured that if I was paying that much to fly to the other side of the world from Australia, I was going to do some travelling afterwards and visit some of the local areas. I got on the internet and googled traditional healers. I had a practice in Adelaide, Australia, doing bodywork and healing modalities and thought that I would love to find a traditional healer while I was overseas.

I could not really find much online. I found the Arvigo website and began getting very interested in this method of healing. However, there were no courses offered during the time that I was to be in Central America. I looked for quite a few weeks on google for other healers. I didn't find anything that felt authentic and real to me. I found different sites where you could pay to go on a trip to different villages to meet healers, but it somehow felt artificial.

After my conference finished, I had a choice of heading through Mexico to the Chiapas region and then down into Guatemala, through Belize, and back to Cancun where my trip began, or I could travel the other way around, down into Belize, through Guatemala and then back into Mexico to Cancun. I did find on the internet a lady who had studied with Rosita Arvigo in San Ignacio, so I decided that because that was where the Ix Chel Farm was I would head there and just explore the area.

I was at the border of Mexico and Belize when I met a young woman who was heading back into Belize after birthing her third baby in the USA. She lived in Belize and was heading home. I asked her if she knew of any traditional healers and she mentioned the name of a lady named Beatrice who she had seen on one occasion. That was all the information she had.

I headed to San Ignacio not really knowing what I would find. I set up home there for about four days and continued to ask around if anyone knew a lady named Beatrice. After a few days, I eventually found someone who had a phone number for her. I tried dialing the number numerous times, but it just kept ringing. In the end, someone took me to her daughter's house and they rang her for me and said that Beatrice would be happy to see me. I took a bus to Santa Familia that afternoon and followed directions to Beatrice's home.

Meeting Beatrice and her family was an experience I shall never forget. As I look back, I grow more and more appreciative of the time I spent with her and her willingness to take me on as a student in her home. I ended up staying four days and four nights, with a two day break in between, followed by another four days and four nights. I slept in a small brick hut just down the hill from her home past her chicken shed and surrounded by jungle leaves.

My first experience of Miss Beatrice was of climbing the white stairs of her home hearing Spanish being spoken out in strong tones above me. I stepped into the kitchen and found two of her daughters, Zena and Jeanette, there with Beatrice, sitting at her kitchen table. Small, dark skinned with black curly hair, she sat before me, a lady full of spirit and eager to share her knowledge. We spoke a little about what she teaches and of her knowledge. She was clear and direct with me, confident of her abilities of passing on knowledge and knowing that her time was limited in doing so.

That night I returned to San Ignacio for a final night before packing my things and moving to Santa Familia. Miss Beatrice and her family were incredibly warm and inviting to me, a stranger in their home. Her daughters, Zena and Jeanette, took great care of me, ensuring that my bed and room were clean and that there was food each day for breakfast, lunch and dinner. Whenever I had to go back to San Ignacio via the swinging bridge over the main river where they lived, she arranged for her son to pick me up and carry me back on the rear of his motor bike over dusty dirt roads and pot holes, avoiding dogs and chickens. On one occasion we were in the face of a torrential downpour. It was hilarious!

While I was there in Beatrice's home, we had sessions each morning and each afternoon. She organized breakfast in the morning, a lunch break in the middle of the day, and a delicious meal in the evening, surrounded by laughter with family members. The food was always freshly cooked and was often a meal of tortillas (which Zena taught me to make), fresh beans and salsa. Sometimes there was fish, fresh fruit, rice and chutneys. She always ensured that the food was to my liking and that there was enough. I am a fish-eating vegetarian, and they always made sure that the food was a balanced, wholesome diet for me.

In the evenings, she would send me back to my little hut dwelling, with black and white copal to burn as the sun went down to keep the energy in the room filled with friendly spirits, ensuring my safety. I was given strict instructions to not go outside at night, so that the wind would not get me and, I would remain safe. I used to spend a little time outside in the dark before going to sleep each night, under the torch light, watching the ants roam around outside the hut. There were thousands of them marching through the jungle, carving little paths and carrying little leaves on their journey. It fascinated me.

One night she prepared a herbal bath for me; another night there was a vaginal steam ready for me to enjoy. Zena took me around the garden and showed me all the healing plants that were there. It was incredible. I wish I could remember them all. One after another, I saw amazing plants surrounding her house with all these healing properties that the Mayas have known about for thousands of years. I began to get jealous of their tropical weather with warm humid days and so much rain! The plants all seemed to be so healthy, abundant and easy to grow. Coming from the desert terrain of South Australia (much like Arizona), I was really appreciating the change in climate and the tropical weather.

Miss Beatrice taught me all about abdominal massage. She taught me the technique and had me practice on women, family and friends that she had invited to come to her home. She had me practice for days until she felt that I had the technique just as she liked it. She would be very strict with me and would get cross if I left something out. She was insistent that the session was never over until the ladies feet had had a proper massage!

Beatrice also taught me some of her Maya spiritual healing methods. This I found fascinating, as she taught me some of their herbs, prayers, pulse reading, throwing stones ceremony, herbal baths and vaginal steams. She was a wealth of information and never once picked up a book or had to reference anything. It was me who was madly trying to write it all down! She taught the traditional way, through words. You could ask her anything about any herb or issue with women, and she would have the answer.

While I was there, I took a break from her teaching and travelled into Guatemala to visit Tikal for a few days. While I was staying there, I was in a room next to an older couple who seemed very disturbed, talking and yelling to themselves, banging on things and generally behaving in ways that actually creeped me out. I was only at this hotel for two nights. Then I headed back to San Ignacio to return to Beatrice's house. The first night I was back with her, I learned why everyone seems to have roosters.

While in Guatemala, I had heard them crying out, every morning and waiting to hear the call back from the dozen other roosters living in the same area. It had become something that had begun to drive me a bit potty! The first night after I returned to Beatrice's house, her rooster started to behave in a very peculiar manner. He was not crying in the morning, but at 2:00 and 3:00 am continuously. It kept me awake for hours. I began to get very angry about it. The next morning, as I sat at the breakfast table falling asleep, I told her what had happened and how angry it had made me and how I just wanted to swear at it and how I was very upset! Miss Beatrice took this very seriously, as this was a strange change in my behavior. She also said that the roosters are there to scare away the evil spirits. According to Miss Beatrice, this was a sign that I had brought something back with me from Tikal and that the rooster was trying to scare it away. I told her about the couple I had stayed next to and this seemed to confirm her suspicions.

Beatrice made me lie on her table. She waved a fresh chicken egg over my being while saying certain prayers silently. She held it on some areas of my body more than others. Afterwards, she broke the egg into a cup and showed me what she said was an evil spirit in the cracked egg yolk. I was directed to take it away from the property and put it into the ground and walk away without looking back. Needless to say, I followed her instructions. I did not feel angry about the rooster anymore. I was not woken again by him crying out at strange hours of the night!

The next day she had me make a pouch filled with special herbs and wood from certain roots and trees. She sent me down to the riverbed with directions and prayers to sew and create the pouch for the herbs. Later she stuffed it full of different plants and a sacred root from a tree called skunk root. This is a very sacred tree to the Maya and valued highly. To access the root, the whole tree must be killed. For this reason, it is used with care and consideration. I was to carry the pouch on me at all times to keep me protected and safe getting home.

Miss Beatrice was a wealth of knowledge and love. Her dedication was inspiring, as was her hard work and perseverance through the thick and thin of her life. Her home was full of joy, love and wonderful family stories. She shared a lot of herself. Miss Beatrice not only taught the historic knowledge of Maya healing, she also shared stories from her life. She was a committed midwife and healer who had this skill from a very young age and spent her life helping others. It was a very warm experience staying with her family. I am sad to know that Miss Beatrice has passed on and is no longer with us to share her knowledge. I trust that her work has spread and her knowledge lives on in her students as they work to support the healing journeys of others the best way they can.

A SHARED LOVE OF PLANTS

By Joy Kemna

On my first adventure to study Maya Healing with Miss Beatrice, I delighted in experiencing the wondrous, healing life force of the plants of the jungle. My excitement was even greater when she shared which plants I could take home with me as leaf cuttings or seeds in order to start as tropical healing houseplants to work with in my northern climate. I remember the reverence with which I collected the nine castor seed pods from the second level staircase outside of Miss Beatrice's kitchen. (Had I known that each pod contained three castor beans, I would have taken many less)! The green treasures that made the journey from Belize have supplied countless spiritual baths and healing compresses for my family and community.

The following year Miss Beatrice came to the United States on a teaching trip. We had the pleasure of making spiritual baths in a garden full of the medicinal herbs of Colorado. Pausing for a moment, from our plant collector's prayer, I looked up to see Miss Beatrice in deep communion with a dandelion. Imagine my surprise when she slowly collected the dandelion seeds and carefully packed them away for her journey home! At this moment I felt the resonance of a gardener's heart, which delights in the seemingly simple yet astoundingly grand gifts from the plants. For Miss Beatrice, tending her garden was truly part of her devotion and was one of the many important tasks that she devoted her life to.

Miss Beatrice worked hard. The parts of her life that I was privileged to witness included countless acts of service to her family, her community, and to the earth. Miss Beatrice wasn't stingy with her gifts as a healer. I remember her telling me the story of how she performed uterine massage on a dog that had just undergone a traumatic delivery of her litter. The plants, the animals, the winds, and the rains were as integral as her connection to humanity and the spiritual world. Miss Beatrice put in the time, with her students, whether it meant working long hours or traveling great distances, to ensure that they understood the Mayan teachings. I remember how she stayed up past midnight to prepare tamales to welcome us and to honor the benevolent spirits. She gave of herself in the most profound ways so that her children would have better futures. In so doing, she passed along teachings that have provided better futures for the rest of us. I still see her in the garden–reverent, awed, and joyful–surrounded by the healing plants.

SLEEP WITH THE ANGELS

By Katherine Silva

Miss Beatrice has always been straight forward with me and right on the mark. She told me not to have any more children after my last child. She said that it would be better that way, but I didn't heed her advice and I purposely conceived, only to have a very traumatic miscarriage.

Now I understand that my womb energy needs to be dedicated to being a healer, a teacher and a writer, not to raising more children. She could see all that in me before I could begin to grasp it.

She once read an egg that another student had passed over my body in a healing workshop and said, "Many people envy you. You need to keep your mouth shut and stop bragging about everything. You are coming into some hard times-not really bad ones-and not for too long-just some hard times."

Although I had wanted to hear something more positive, she was right as usual. Hard times came just around the corner! Ever since I listened to her warning, I have learned how to pull my energy in and contain who I am and what I do with a little more dignity. I learned to honor myself by not telling everybody everything and by not scattering my precious energy all over the place. My life has changed for the better because of that. I also did come into a rough period right after she told me that but as she said, it wasn't too bad, and it has passed.

The healings and advice Miss Beatrice has given all four members of my immediate family have been invaluable, life changing and profound. Not a day goes by that I do not incorporate something of what she has taught me into my daily life. In our home, we burn copal, light our candles, say our prayers and do our plant spirit baths and bajos and ceremonies. I am careful not to chill my feet on cold floors or go out in a cold draft after a warm bath or shower.

I remember the first time I actually remembered to wear a towel on my wet head after taking a shower in her tropical home and she said, *"You are finally getting it!"*

I drink my herbal teas, especially our beloved hibiscus tea, every day and serve them to my family. At home, we enjoy many of her scrumptious recipes. We avoid cold drinks and ice cold food. I eat lots of greens. I have rue and rosemary growing right outside each of my doorways and each summer I grow a big crop of basil and marigolds. I massage my uterus everyday when I am not menstruating.

Before conceiving our son, I spent months preparing for pregnancy with womb massage, prayers and womb steams. We conceived on the very first try. When he turned breech around thirty-eight weeks, the womb massage encouraged my breech baby into the perfect position! I incorporated all that Miss Beatrice had taught me about the Maya way of birthing into my son's home birth, such as eating okra during the last month before the birth to make the baby "slide out," drinking the special labor tea, staying up right during labor and honoring Ix chel. The birth was easy and very quick and very blessed. My son, I believe, is a gift from the beneficial Maya Spirits. Some of his antics have made Miss Beatrice break into gales of laughter. She likes to recount the time he interrupted a ceremony at age two with hands on hips saying, *"What* are you *doing?"*

Every ceremony I have been blessed to attend with Miss Beatrice has brought profound blessings to my life. Every single thing I ever asked for at her ceremonies has come true, including finding my husband, conceiving my son and finishing this book.

We wear amulets and tuck some basil, marigold or rue in our pockets when we go out and about for protection. I do frequent plant spirit baths and egg clearings on my children that serve to switch any cranky mood immediately to joy and laughter and presence. The prayers and practices really work. I don't know what I would do without them!

I have incorporated these teachings into my mothering, my massage practice, my herbal practice, my doula services and my midwifery apprenticeships. I am grateful everyday that I have this connection and that these ways have been preserved. They have benefited countless women and babies in a profound and lasting way.

A year before Miss Beatrice passed away, I had a dream that she was running away from me very fast. She was wearing her white primicia dress and looked to be very young. I chased her. She ran all the way to the river. She was about to dive in. I called to her, "Be careful don't fall in the river!" I caught up with her and grabbed her to protect her from falling in. She grabbed my arm and pulled me in with her. We dove deep into the river. I rode on her back. We swam along the surface of the river for many miles. I looked down. Arranged on the river bottom in perfect formation were nine skeletons, one after the other. The bones were glowing white. We swam over them, one after the other, until she pulled me up out of the water. We splashed to the surface and she pointed to a cave in the hillside. My body rose up out of the water, and with my arm outstretched, I flew up to the cave. Inside the cave was a beautiful blue shining stone. I woke up just as I touched it.

When I told her about the dream, she sounded shocked and said, "Katrin, what kind of dream is *that?"*

She then told me that the dream meant that I would find that blue stone and a matching one for her, which I did. She has hers and I have mine. It is a stone to assist with dream visions. As the years pass, the entire meaning of this dream is becoming more and more clear to me.

A few weeks before Miss Beatrice passed away, I had a dream in which she came to me and said, "Don't worry Katrin. I am not going away, I am only coming closer."

I awoke with an immense sense of peace. This dream has certainly turned out to be true!

The last time I spoke with Miss Beatrice, she reprimanded, "Katrin! The only thing wrong with you is that you are working too much! I know you want to do this *midwife thing* but you have to take care of yourself!"

I quickly changed the subject by asking her what the weather was like in Belize and she said, "It's REALLY hot.... but my hibiscus tea is soooooooo refreshing."

At the end of our conversation, she told me that she would always love me and in her classic way of saying goodnight, she said so softly, "Sleep with the angels, Katrin." And I do.

BELIZE'S SHINING LIGHT: MISS BEATRICE WAIGHT

By: Laura Lozano

My road to Belize was met with hesitation. After a nature walk with a group of herb lovers, a friend told me that she was taking a group of women to Belize to learn self-care Maya Abdominal Massage. My initial reaction was, "Thank you for asking and no thank you." I didn't have a uterus, so why would I want to learn self care Maya Abdominal massage? She asked me to think about it and I said I would. A few days later, I was talking with a friend about it and she looked me straight in the eyes and said, "You have to go." It was in that moment that I felt it too. I felt goose bumps and knew I needed to go to Belize. I wasn't completely clear why it was important for me to take the trip but I knew it was right for me. I let my friend know that I was going and began preparing for my trip to Belize.

In February, 2000, I traveled with a group of women to Belize to learn self-care Maya Abdominal Massage. With a brand new passport in hand, I began a journey that still continues today. Not only was Belize the place I received the first stamp in my passport, it was also the place where I met the woman who became my mentor, my teacher and my friend, Miss Beatrice Waight.

The moment I met Miss Beatrice, I felt as if I had known her all my life. I was immediately drawn to her and for me if felt like two old friends coming together after years of not seeing one another. It was comfortable. I spent hours talking with her and finding out about her life. She shared stories and her wonderful laugh. Miss Beatrice had a laugh that was infectious and an ageless wisdom that she shared with reverence and grace. I learned so much from Miss Beatrice while in Belize. I didn't want to leave.

A few months later, I was told that Miss Beatrice was coming to Colorado and I was asked to host her while she was in Denver. I immediately said, "yes!" and my heart jumped with joy.

Being with Miss Beatrice was a gift. She became part of the family. And my children fell in love with her. It was as if their favorite Aunt had come to visit us. Miss Beatrice was no stranger to working hard and I would watch her as she worked with the people who came to see her. It was an honor to watch her work and I learned so much. She was a powerful and passionate healer. She would speak the truth and her faith was strong. In fact, she would often say to me, "Laura, you have to have faith, you have to believe, if you don't, it won't work." She spoke of the importance of intention. She taught me to be clear about my intentions when asking the Spirits for help. She said that if I wanted to have better communication with the Spirits it was important to pray more, be humble and give gratitude to them through offerings on my altar. Miss Beatrice made it very clear that, "If you want to be a spiritual healer, you must believe in a deity or you can't heal." She also emphasized that it is very important to work with your plant ally when working with people. Miss Beatrice loved her plants and they loved her. I watched, I asked questions and I learned from her.

Miss Beatrice loved her coffee, chocolate and shopping. In fact, when she would visit Colorado, we would frequent a number of shops and botanica's. She loved shopping for her children and surprising them with gifts when she returned home. She had a favorite restaurant near my house and the owner fell in love with her. That happened wherever we went. Miss Beatrice had a presence about her that was welcoming and inviting. She loved life and the adventure. One night while in Denver, Miss Beatrice had a craving for tamales, so we went on a hunt for tamales. We found a place that sold tamales to go and there was a gentleman who was quite smitten with Miss Beatrice. He didn't want her to leave and she handled the situation with grace. She had fun and we laughed all the way home. I loved Miss Beatrice's laugh. Her laugh lit up a room and made you laugh along with her. The night before Miss Beatrice would leave to go back home to Belize, we would begin the task of packing. I would pack her bags and fit everything she bought and more into her bags. She called it magic packing and that became the term we used whenever we packed her bags.

Miss Beatrice's faith was strong. She believed in what she did and knew that if she asked for help, help would be given. To watch her work was like watching an artist create a masterpiece. I was blessed to be in Miss Beatrice's first apprenticeship group in the United States. There were seven of us and with careful instruction from Miss Beatrice we were entrusted with the sacred Maya traditions. Learning from Miss Beatrice was a gift, one that I am most grateful for to this day.

Miss Beatrice was human and endured many challenges in her life. Sometime they got to her, however, she had faith and trusted that it would work out, and it did. She was a strong woman, a caring woman and a sassy woman. I loved that about her. She had passion and was full of spirit. She was honest and direct and sometimes it was hard to hear what she had to say, however, it was the truth. She challenged you to be better than you believed yourself to be and gave you ways in which to heal your wounds. A healing with Miss Beatrice was an amazing experience. She offered you beautiful rituals, prayers and healing; all of which were medicine for your soul. Your job was to be willing to do what was asked of you and let go.

Attending a Primicia given by Miss Beatrice was a blessing. Each participant felt the sacred wisdom and honor in each and every step. The power in each word sung to each direction, the meditative moment in which the veil was lifted and you were in communion with spirit and your soul. These special evenings would end with the joy of sharing a feast with friends and celebrating the healing that had just taken place. It was a magical and sacred journey that was an honor and a privilege to behold.

Miss Beatrice loved people and I feel blessed to have been part of her life. Her wisdom is not lost. It is carried on by those who met her, who loved her and who were her students. We carry the mantle of wisdom she gave to us and with her spirit lodged into my heart, I move forward gratefully carrying her torch of healing wisdom. Thank you Miss Beatrice for shining your light upon our world! I love you and I miss you.

FOREVER IN MY HEART

By Zena Waight

My name is Zena Waight. I am the youngest daughter of Miss Beatrice. My mother was the most precious treasure I possessed; she was my super mom, always full of life wherever she went. Even though she went through hardships in life, she never gave up on her nine children; making everybody laugh with her stories and jokes and giving us positive advice and energy. My mom is my biggest inspiration as a women and I am thankful to have had her as a mother. Every day I pray and look up to the stars at night and feel my mom smiling down upon me like she always did.

Her memories will forever be in my heart. I miss you Mom!

Love, your "nene"

Herbal Ally Number Eight: Castor

Ricinus communis

Miss Beatrice with castor leaf (photo by Janet Caspers)

Another name for castor plant is Palma Christi, or "the hand of Christ." The big beautiful leaf is shaped like a hand and is so helpful in many ways. Castor can be placed on an aching head to cool and soothe it in quick time. Used in the same way, castor leaf can bring down a fever. A castor leaf rubbed with olive oil and placed on a benign breast lump for five days, with prayers, will make the lump disappear. For constipation, we take the castor stem, put castor oil on it and insert it like a suppository. After ten minutes, the person will not be constipated anymore! One ounce of castor oil can also be mixed in a cup of warm water and taken internally to flush the bowels and the uterus after childbirth. I have a big beautiful castor tree right out my kitchen window. I love it and use it a lot!

Chapter Nine: Journey to the White Light

Miss Beatrice in her healing hut with a corn husk angel (Photo By Monica Juchum)

"Mayas believe that not until after three days after death, does someone actually know that they are dead. On the first day, they don't know. On the second day, they still don't know. On the third day, they kind of wake up and all of a sudden they know what has happened. That is when their new journey begins."

Traditionally, Maya people die at home. Close relatives dress the person in special clothes after they die. For women, the chosen clothes are always white. The clothes might be adorned with lace and all embroidered, but anything that is new and white is sufficient. We dress the men in a white shirt and black pants. When someone dies in our culture, we have a wake. The villagers and all the friends that knew the deceased attend the wake and bring something to share like coffee, bread, little cakes, soft drinks, rice, beans and tamales. We all share food and songs and prayers for the whole night. Sometimes the wake lasts for two whole days.

On the morning of the funeral, the family goes to the cemetery and digs a grave or makes a tomb in the ground. They dig a hole and put the body inside it in a coffin. The dead person is always treated with white lime in the anus and arm pits so that he doesn't smell bad.

Before the group leaves for the cemetery, if the one who died was married or was a child or an adolescent, the person who knew them the best stays behind at the house and calls the name of the dead person nine times. Then they say to her, "Today is your last day in this house. From now on you will live in another house, which is in the cemetery. You don't belong in this house again!"

We do this so that the person who has died won't stay in the house and haunt the other family members. We are asking the spirit to go and dwell where the body is because the spirit has a long journey ahead of him and we don't want him hanging around the house bothering the living! We need to tell the deceased to go because Mayas believe that not until after three days after death, do they actually know that they are dead. On the first day, they don't know. On the second day, they still don't know. On the third day, they kind of wake up and all of a sudden they know what has happened. That is when their new journey begins.

After the funeral, everyone who attended comes back home. They wash their hands in a basin of marigolds. This holy water, infused with marigold blossoms, gathers the entities and anything unwanted and takes it all out of the person and into that water. The water then can be given to the earth away from the house.

The family fills the bed where the person died with gumbolimbo branches and offers a prayer everyday to ask the angels to take care of the dead person's spirit and to lead him to the white light. The family makes a little altar with a white candle, a glass of water and some white, amber or grainy copal, to light the way for the journeying spirit. The main journey to the light takes nine days. It is an initiation, a transformation and a time of reflection. It is an opportunity to fully receive the lessons the person had while on earth.

A little ceremony you can do when someone dies is to put a white candle on an altar and make a little dough with the initials of the person who has died and give him what he liked to drink and offer a little prayer for him to go to the white light. For nine days, burn yellow or grainy or white copal or white sage but not the black copal because that chases him away without help. The spirit really leaves after three days because they find themselves knowing what has happened to them. Burn your little white candle for nine days to help them find their way to the white light. In life, they may not have had time to repent, so the nine days gives them enough time to repent and let go. After nine days, they find their path and go into the white light.

On the ninth day, the soul is all the way in the white light so we put all his clothes, shoes and belongings out of the house and we celebrate with a memorial service. We prepare food and drinks, especially the kind that the deceased liked to eat and drink. The immediate family and friends gather in the home. A mass at church is also done. We put out pictures on the altar of the deceased and give thanks to God, our creator, for the life of our loved one. At the end of the service, everyone eats, drinks, and remains on the premises until the wee hours of the morning!

At the one year anniversary of our loved one's death, during the afternoon, the immediate family members visit the grave. We bring flowers to the grave and light candles. The family also organizes church services in the evening and religious music is played. We make an altar with pictures of the deceased, pretty flowers with white candles. At the end of the gathering we all eat together.

Our family honors our loved ones for our whole lives. We call the first and second of November All Saint's Day. In Spanish, it is called *Día de los muertos.* On the first of November, we remember the young babies and children who died, and on the second, we remember the adults. We often go to church on these days and there is a special mass.

At home, we do our own services. We make an altar with photos of our ancestors. On the altar, we have white flowers and white candles, as well as favorite food of the Maya. We make *escabeche,* also known as onion soup, and black soup, known as *chirmole.* We also have bread, pastries and favorite puddings. We give thanks for the existence of our loved ones and commemorate their lives. At the prayer service, we each voluntarily pray in a slow, quiet voice or silently in our minds. Then food is eaten and shared among family members, always with the deceased in mind.

At the one year anniversary of a loved one's death, during the afternoon, the immediate family members visit the grave. We bring flowers to the grave and light candles. The family also organizes church services in the evening and religious music is played. We make an altar with the pictures of the deceased and have pretty flowers with white candles. We all eat together.

When Mayas bury their dead, they collect all his belongings and bury them with him or give them away to charity, depending on the requests the deceased left behind. They leave one set of clothes to remember him by. Everything else is put on the tomb to go with the person unless the person said he wanted a particular thing given to somebody. Sometimes they have a chance to make their wishes before they die. If not it all gets buried with them. It's a bad omen to use things that belonged to them if they didn't give you permission.

Now I will tell you a little story so I can make my point very clear. A lady I once knew had a big container and she said, "Miss Waight, do you want these little pots and kitchen utensils, and some books?"

I said, "Sure" and I went there and I gathered some little balls that make noises and I found a special angel.

She said that I could have it. She was moving and so she said, "If you could use it, just help yourself."

I was attracted to the angel, so I took it and I set it by my bed where I was sleeping. In the night, I did my prayers and in my dream that night, I saw a tall guy, kind of slim, with a dark complexion and wavy hair.

I dreamt that he said, "Okay, Miss Waight, you could take anything from the container, but not that little angel. The angel you cannot take because it belongs to my daughter and you cannot take it anywhere. That is my special angel, so I don't want anybody to have it except my daughter."

In the morning I told the lady about the dream vision and asked her to tell me if it was true. I said, "I dreamed that your father came to me and he was not mad, he was smiling, but he said that I could have everything from the container except the angel because he doesn't want that angel to go far away from you."

She cried and cried and sobbed, "Those were the same words he told me, but I just thought if you wanted the angel you could have it."

I refused, saying, "No, he doesn't want me to have it. He wants you to have it!"

So, she agreed, "Okay then!"

I handed her the angel and explained, "As much as I like it, I cannot take it, because it belongs to you."

I never met him in life, but I could see and hear him in my dream. So you see, dreams are powerful! I was afraid to take the box of other things, but she said it was okay. The whole thing scared me a little.

We Mayas try our best to respect the dead. If someone was cremated, we do not keep the ashes too long in the house. If the person liked a special place in nature, like the creek or the sea, we put the ashes there. We figure out what she liked in life and it is there that we throw the ashes. If we keep the ashes too long we believe that we are punishing that soul, who simply needs to go to the white light.

I will now tell you a story about what can happen if the soul of a deceased person gets no ceremony. One time, I went to Massachusetts. I was staying downstairs in a big house with Miss Hortense Robinson who is a midwife from Belize. Everyone else was staying upstairs. After ten o' clock at night, there was someone crying outside on the porch and a big commotion going on and people murmuring and trying to open the door. I wasn't gonna open it because I was in a strange place and I didn't know what was happening. I was going to let the owner come down and check it out because I didn't know the neighborhood. It went on like that for almost an hour, so I kind of went and poked Hortense and asked her if she could hear the noise. She said it was a bear, which scared me even more, but I knew it wasn't a bear. It sounded like the voice of a little girl shouting for help. I said maybe she is being beaten by her parents or something bad. In the morning, I told the owner that someone was making a lot of noise by the door last night. He said it was the bear cubs. I told him it didn't sound like bear cubs! I said it sounded like human beings. It was like a little girl crying out for help.

"Oh", he laughed, "That's the little ghost girl! She was attacked by a bull on the porch and was killed right there and from that day on she comes every night. That's why the previous owner sold this farm and went somewhere else, because it tormented him every night. I asked him what he thought about it. "We just don't pay any mind," he replied.

He asked me what I could do, and I said, "That child needs a ceremony to send her to the white light. She is tormenting herself right there, because she can't go anywhere! We could do a ceremony tonight to send her on her way."

He agreed to host the ceremony. We didn't know her name but we knew it was a little girl. We put a glass of water and some white candles on an altar for her. There were six of us, each offering a candle on the altar. We offered chocolate and some apples and little foods that a little girl would like. The table was full with little offerings; then I started the prayer and the others did some prayers as well. It took us an hour and a half.

All of a sudden, we heard a big group of people running away. I told them she was going away and going to the light. We stayed in that house a week longer and no more commotion happened. There was no more crying. I think she had made the noise so we could hear her and help her pass to the white light. We gathered white flowers and white roses and pretty colorful flowers because kids like colors so much, and that showed her that we were there to help.

During the ceremony, we burnt the white copal inside so she would come and feast and think that we are her friends. She never came back after that, so she really needed that help and she got it.

Maya Plant Ally Number Nine: Marigold

Tagetes Erecta

Miss Beatrice and Marigolds (Photo By Sandra Lory)

Marigold is one of my most important plant allies. In Maya we call it, *Ix ta pul*, the golden headed goddess. I use it in combination with basil and rue to make plant spirit baths. I love to use marigold when I whisper prayers into pulses. Marigold also wards off negative energy. That is why we never use marigolds in ceremonies which invoke benevolent spirits, invite good energy or offer gratitude. It protects us when we tuck a marigold flower in our pockets when we go to town. When my friends come to visit me in Belize, they always bring me marigold seeds. I plant them in my garden and all over my yard. The more marigolds, the merrier!

The Afterward by Katherine Silva

Our beloved Miss Beatrice transitioned out of her physical body on October 3rd, 2010, after a peaceful morning surrounded by her loving daughters. Many of us all over the country offered the candles, the water, the copal and the prayers that are customary in Maya culture to support the one who is crossing over to have a good and complete journey. I had a series of dream visions beginning after the ninth day of these death rituals.

The main dream vision repeated three times and unfolds as follows. I am rocking in Miss Beatrice's hammock in her beautiful round healing hut. I am rocking, rocking, rocking with the jungle breeze. I open my eyes and I see Miss Beatrice standing over me. She smiles and says, "Katrin, if you really want to know what we Mayas do, all the way, go to the Yucatan."

She then puts nine thorns in my abdomen in the shape of a cross. That's where the dream ends. The identical dream came to me a second time. The third time I had this dream all was the same, but instead of Miss Beatrice, it was her father speaking to me and putting the Maya acupuncture thorn formation in my belly.

I knew better than to ignore a message like this but did not see how I would pull off this kind off trip any time soon. I prayed for help and my prayers were answered with another series of intense and vivid dreams, all involving the ancient Maya city of Chichen Itza. When I looked on a map I saw that Chichen Itza is twenty miles from Vallodolid, the place from which Miss Beatrice's Grandparents migrated to Belize. Amazed, I realized that Miss Beatrice and her father were guiding me, through dreams, to their ancestral home. Within a month, miraculously, the funds appeared, and so we went.

I found, in the Yucatan, the Maya ways that Miss Beatrice so lovingly passed onto her students, being practiced in full swing and very much intact, alive and well. Still being practiced, among the Yucateca Maya people near Chichen Itza are; the Hetz Mec Ceremony, the primicia, the abdominal massage, the Maya spiritual healing, the same plant uses, the rituals of asking permission to the Gods and plant spirits before harvesting plants, the Maya birthing practices, the worship of *Ix chel* and the deep and clear path of faith that Miss Beatrice walked. Most encouraging of all, I found the *Ix'men* (pronounced eesh men) and the *H'men* (pronounced accchmen), like Miss Beatrice's father, who are the Maya priests and priestesses who take responsibility for caring for their communities, not only spiritually with blessings, healings and ceremonies, but who also know the ins and outs of the jungle, the names and uses of hundreds of plants and how each one is harvested, when it is best harvested and with which prayers. They use these powerful plants to heal and console the members of their communities, young and old.

Through my dreams, Miss Beatrice lead me to a Yucateca Maya *H'men* who comes from a long line of *H'men*, on the male side and a long line of midwives on the female side.

He listened patiently while I told him, in my flailing Spanish, of Miss Beatrice and of my dreams. Then he confidently announced, "We will have a ceremony in the morning. I will let the Gods tell me what I am supposed to teach you."

The next morning he conducted a traditional Maya ceremony for me in the sacred ceremonial grounds, asking for this guidance. After the ceremony, he told me that the Gods told him that I had been guided to him because my training had started and gone well with Miss Beatrice but that it was never finished. He immediately started training me in more detail on how to conduct a Maya ceremony. He trained me to conduct Maya ceremonies such as the primicia (called *Ho Che* in Yucateca Maya) and the Ix chel ceremonies. He also worked with me for many days, teaching me more about the sacred Maya way of harvesting and working with jungle medicinal plants.

While I was staying there, I had a dream vision of Miss Beatrice telling me, "This man is like my father. I would have given anything to be trained directly as you are being trained right now. Pay attention, listen well and bring these teachings back to my students and my children."

Our blessed time together culminated in a beautiful initiation ceremony. At the end of the ceremony, he put his forehead to mine and told me in Spanish that from this day on begins my work with people.

As the ceremony was ending, an entire crowd of tourists with cameras showed up out of nowhere. I was kind of shocked, but I had remembered that he had taught me that whatever happens during a ceremony is a message from the Gods. I had to leave swiftly after the ceremony to travel on with my family, so I didn't get a chance to ask my new teacher what he thought about the invasion of *los touristas.*

That night, sleeping by the ocean, I had a dream of my new teacher right up close to my face. I asked him what he thought about the tourists in the ceremony.

He comforted me by saying in Spanish, "They are symbolizing all the people of the world who are hungry for this work."

My experience in the Yucatan awoke within me an even deeper appreciation of the bright, strong, shining soul of Miss Beatrice, who through all obstacles, learned, carried, held, protected, practiced and taught these precious ancient teachings intact and in full force. She carried this torch through a difficult childhood and young womanhood, through child bearing and child rearing, through the disintegration of her culture around her and through her own death, where she carries them still and shares them with us in our dreams.

Ann Drucker after a primicia (photo by Monica Juchum)

One night, in a dream, a large angel came to me. She explained that she would be the one to lift Miss Beatrice up by the armpits and stretch out her shoulders and arms like the outspread wings of a large bird. She said that Miss Beatrice would fly away with her in peace.

I met Miss Beatrice in February, 2000, in Belize. She immediately welcomed us into the magical world of Maya healing. Over the next ten years, she traveled many times to be with us in Colorado, and we to study with her in the beautiful village of Santa Familia on the Mopan River.

It's been an incredible gift for me to know Miss Beatrice and her nine wonderful children and fourteen precious grandchildren. They are full of love, laughter and joy. Their kitchen buzzes with fresh warm tortillas, soups, salsas and kindness. We have become extended family to one another.

In the last days of September, 2010, I dreamt that Miss Beatrice was giving me a new teaching, but not in English or Spanish. When I looked to her daughters to translate, they helplessly shrugged their shoulders. I woke knowing it was again time for me to visit her.

Marian Rose and I traveled to her home. Her spirit shone brightly; her body was tired. We feasted with Miss Beatrice and tended to her Maya healing ways. We roasted fresh rabbit on her fire heart. We honored the first year anniversary of her mother's death with homemade tamales and her favorite chicken soup. We massaged healing oils into her sore muscles, burned copal with prayers, bathed and brushed her with plants. We dreamt of angels nightly. On the way to the market one morning, we saw a pair of condors sitting in a tree with their wings fully extended.

One night, in a dream, a large angel came to me. She explained that she would be the one to lift Miss Beatrice up by the armpits and stretch out her shoulders and arms like the outspread wings of a large bird. She said that Miss Beatrice would fly away with her in peace. "Might be quiet, or may be noisy," she said with a mischievous smile.

When I shared this dream with Miss Beatrice, she said to me, "I'm not afraid to die. We all die, every one of us here alive on this earth. I have made my peace. I know I will be taken care of by the angels."

Six days after returning from Belize, on Sunday, October third, I received a tearful phone call from Miss Beatrice's daughters. She had died peacefully, after spending a delightful morning with her daughters at her bedside.

Miss Beatrice, how we treasure and love you. You are forever in our hearts. Now you are with your beloved angels, flying with your big wings, exploring new realms. You have truly left a beautiful legacy here on earth. Thank you.

Miss Beatrice's smile (Photo by Laura Lozano)

Plant Index

Allspice: Pimenta Dioica

Aloe Vera: A. Barbadensis

Amaranth: Amaranthus Dubius

Balsam copaiba: Myroxylon Balsamum

Black mango tree: Mangifera Sp.

Blue vervain: Stachytarpheta Cayennsis

Calendula: Calendula Officinalis

Cassava: Manihot Esculenta

Castor: Rinicus Communis

Catnip: Nepeta Cataria

Cayenne: Capsicum Annuum

Chamomile: martricaria recutita

Chaya: Cnidosculus

Chocolate: Theobroma Cacao

Cinnamon: Cinnamonum Verum

Coconut: Cocos Nucifera L.

Culantro: eryngium Foetidum

Cultivated Basil: Ocimum basilicum

Fat Oregano: Lippia Graveolens

Fennel: Foeniculum Vulgare

Garlic: Allium Sativum

Ginger: Zingiber Officinalis

Guava Leaf: Psidium Guajava

Gumbolimbo: Bursera Simaruba

Hibiscus: Hibiscus Rosa- Sinensis

John Charles: Hyptis Verticillata

Kava Kava: Piper Methysticum

Lavender: Lavendula Augustifolia

Life Everlasting: Kalanchoe Pinnata

Lipstick Tree (Annatto): Bixa Orellana L.

Manvine: Securidaca diversifolia

Marigold: Tagetes Erecta L. Officinalis

Moses in The Cradle: Tradescantia Spathecea

Neroli: Citrus Sinensis

Nopales: Opuntia Cochenillifera

Okra: Abelmoschus Esculentus

Passionflower: Passiflora incarnate

Peppermint: Menta Piperita

Red China Root: Smilax Sp

Rose: Rosa Chinensis

Rosemary: Rosmarinus Officinalis

Rue: Ruta graveolens

Skullcap: Scutellaria Lateriflora

St. Johns Wort: Hypericum Perforatum

Thin Oregano: Oregano Castillo

Valerian: Valeriana Officinalis

White Copal: Protium Copal

Wild Belizean Basil: Ocimum campechianmum

Wild cucumber flower: Methothria

Wild Yam: Dioscorea Sp.

Yarrow: Achillea Millefolium

RESOURCES

To order more copies of this book or for more information on Miss Beatrice, her life and her teachings, go to www.missbeatrice waight.com. For discounted bulk book orders email katsilva9@aol.com. To learn about Maya cooking, Belizean herbs and Maya culture in Belize, contact marlyn_waight@hotmail.com or twaight@ymail.com

Through hard work, dedication and generous donations of friends and colleagues, Miss Beatrice sent seven of her older children to college. They are in the world now holding respected service positions. She didn't get a chance to send her oldest daughter or her youngest daughter or any of her grandchildren to college before she passed over. A fund has been created to send these Maya youth to college. To donate to The Maya Youth Educational Fund: contact Ann Drucker at drucker.ann@gmail.com. Buying this book also benefits this fund. Thank you!

Katherine Silva practices massage therapy, intuitive healing and many of the practices described in this book. She also teaches workshops and conducts ceremonies. For more information on the work of Katherine Silva go to www.hoodriverhealing.com email katsilva9@aol.com

Ann Drucker offers herbal, shamanic and Maya-inspired healing sessions. She also teaches workshops and leads ceremonies.
For more information on Ann's work go to www.herbalhealingarts.com

For more information on the work of Rosemary Gladstar herbalist and author extraordinaire go to www.sagemountain.com

For more information on the work of Rosita Arvigo, her books, her workshops, her institute and her humanitarian projects, go to www.arvigotherapy.com

WHAT HAPPENED TO THE ANCIENT MAYA? THEY ARE STILL HERE!

There are about seven million Mayas living in Mexico, Belize, Honduras, Guatemala and El Salvador. Many of these modern Maya still speak Maya as their primary language and still practice their traditional ways of life and spirituality.

Miss Beatrice and her ancestors have made it very clear to several of us that they are still here as well. If you open your heart to them and humbly ask for their presence in your life, they will answer you. They are available for guidance and healing to all who sincerely call on them.

In our dreams they are whispering that it is time for the people of the earth to remember their ancestors and to acknowledge the unseen benevolent forces all around us. They are urging us to reconnect with the healing power of the natural world by treating the earth and her elements with humble care and respect. They are reminding us to ask permission from the guardians of nature before harvesting, digging, building and taking. They are urging us to revive the ancient sacred practices of healing, ceremony, dreaming and plant medicine. They are showing us that in order to hear spiritual guidance and receive spiritual support, we need to take time to slow down and just be. They are inviting us to imagine our bodies being filled with white light so that we can connect with the benevolent divine forces and hear the voices of the ancient ones.

They are telling us that these sacred ways are alive in all of us. Each of us can find our own way to reconnect with our true selves and the benevolent forces all around us. We don't have to be Mayas to do it. We can find the ways of reconnecting that are most passionate in our hearts. They are inviting us to come home to ourselves.

Miss Beatrice's Mother, Dominga Torres and Father, Alejandro Torres